Polyamide –
The Evolution of a Synthetic Membrane for Renal Therapy

Contributions to Nephrology

Vol. 96

KARGER

Basel · München · Paris · London · New York · New Delhi · Bangkok · Singapore · Tokyo · Sydney

Polyamide –
The Evolution of a Synthetic Membrane for Renal Therapy

Volume Editors
S. Shaldon, Nîmes
K.M. Koch, Hannover

64 figures and 14 tables, 1992

KARGER

Basel · München · Paris · London · New York · New Delhi · Bangkok · Singapore · Tokyo · Sydney

Contributions to Nephrology

Library of Congress Cataloging-in-Publication Data
Polyamide – the evolution of a synthetic membrane for renal therapy /
volume editors, S. Shaldon, K.M. Koch.
(Contributions to nephrology; vol. 96)
Includes bibliographical references and index. (alk. paper)
1. Hemodialysis. 2. Polyamide membranes. 3. Hemofiltration.
I. Shaldon, Stanley. II. Koch, K.M. III. Series: Contributions to nephrology; v. 96.
[DNLM: 1. Hemofiltration. 2. Kidney Failure, Chronic – therapy.
3. Membranes, Artificial. 4. Nylons. W1 C0778UN v. 96]
RC901.7.H45P65 1992 617.4'61059 – dc20
ISBN 3-8055-5500-8

Drug Dosage

The authors and the publisher have exerted every effort to ensure that drug selection and dosage set forth in this text are in accord with current recommendations and practice at the time of publication. However, in view of ongoing research, changes in government regulations, and the constant flow of information relating to drug therapy and drug reactions, the reader is urged to check the package insert for each drug for any change in indications and dosage and for added warnings and precautions. This is particularly important when the recommended agent is a new and/or infrequently employed drug.

Printed in Switzerland on acid-free paper by Thür AG Offsetdruck, Pratteln
ISBN 3-8055-5500-8

Contents

Preface

Hemodialysis has been available for over 30 years as a treatment for end-stage renal disease. Until 20 years ago, cellulosic membranes were used exclusively, and they still represent the majority of membranes used in hemodialysis worldwide. In the mid-1970s, synthetic membranes became available, and the classic observation by Craddock in 1977 on complement activation by regenerated cellulosic membranes gave birth to the concept of biocompatibility. At just this time, sheet polyamide was being tested as a hemofiltration membrane. It was hydrophobic and strong and was subsequently manufactured in capillary form for hemofiltration. It was also used for water treatment particularly in on-line systems for replacement fluid production in hemofiltration. However, as is often the case in medicine the panacea of the era, *hemofiltration* did not fulfill its expectations and meanwhile hemodialysis was improving. With the discovery by Geyjo in 1985 that β_2-microglobulin was the amyloidogenic precursor of AB-amyloidosis (dialysis-related amyloidosis), the scene was set for a rational development of the polyamide membrane from its original state.

This book contains a series of contributions describing the evolution of the polyamide membrane, both in terms of membrane characteristics and clinical applications. The chapter on 'basic features of the polyamide membranes' recounts in detail the progress that chemical engineering has made to enable the chemist to tailor a membrane exactly to a physician's needs once these are defined. Thus, three membranes emerge: the original hydrophobic high flux polyamide now used only for ultrafilters, and a new, modified, hydrophilic membrane made in two forms. The new membrane is a co-polymer with PVP (polyvinylpyrrolidone) rendering it significantly more hydrophilic. It is spun in one form for hemofilters to optimise shear rate and reduce protein polarization, and in another form for high-flux hemodialysis and hemodiafiltration to optimise small molecule clearance by diffusion. Both forms of the modified membrane provide significant clearance of β_2-microglobulin.

The chapter on 'biocompatibility' of the polyamide membranes explains at the chemical and molecular biology level why the co-polymer has enhanced biocompatibility features. However, it offers a word of caution on the near impossible task of simulating the endothelial cell lining of blood vessels in the body. Again, chemical creativity provided a mechanism by which the original, hydrophobic polyamide membrane, although already highly biocompatible, was improved in biocompatibility as well as in diffusive permeability. The concept of microdomains of hydrophilicity in a sea of hydrophobicity is described and explains the improved clinical performance of the new modified polyamide membrane.

The section on pyrogen retention emphasises the efficiency of the parent polyamide as an ultrafilter for production of sterile dialysate as well as on-line sterile replacement fluid.

The need to keep the original membrane for ultrafiltration is clearly illustrated by some experiments suggesting that the microdomains of hydrophilicity reduce somewhat the total capacity of the device to remove pyrogen. A novel theory as to the mechanism of pyrogen removal by synthetic membranes is also presented, which relates to the presence of double unsaturated bonds in the chemical structure of the polyamide polymer as well as its hydrophobicity.

The section on therapeutic applications reviews the experience from 15 years of hemofiltration and evaluates the most recent clinical mode of this therapy – predilution. The basic chapter on hemodiafiltration describes the theoretical limitations governing ultrafiltration rate and small solute clearance, giving guidelines for the clinician. The theory is supported by clinical experience in the papers dealing with the hydrophilic polyamide membrane used in hemodiafiltration as well as in high-flux hemodialysis. The section on continuous therapies describes the clinical experience with polyamide as a membrane for continuous arteriovenous hemofiltration in adults as well as neonates. The diffusive properties of the modified polyamide membrane also make it well suited for continuous hemodiafiltration or hemodialysis.

The future is addressed in the hope that the modified polyamide membrane will help in treating dialysis-related amyloidosis by removing β_2-

microglobulin and a new inflammatory messenger for neutrophils, granulocyte inhibitory protein. The noninflammatory nature of the membrane and its protector effect against dialysate pyrogen, stimulating the patient's acute-phase response, are stressed.

Finally, a prophecy should be made regarding the potential of hemofiltration, the original reason for developing the polyamide membrane to close the circle of evolution. Clinical hemofiltration with bicarbonate as the buffer justifies the confidence that this mode of therapy will see a renaissance in the future, now that doubts regarding urea as a surrogate for uremic toxicity are again being voiced, and a new era of middle molecular toxicity in uremia seems imminent.

Stanley Shaldon
Karl M. Koch

Membrane Characteristics

Shaldon S, Koch KM (eds): Polyamide – The Evolution of a Synthetic Membrane for Renal Therapy. Contrib Nephrol. Basel, Karger, 1992, vol 96, pp 1–25

Basic Features of the Polyamide Membranes

H. Göhl[a], *R. Buck*[a], *H. Strathmann*[b, 1]

[a] Gambro Dialysatoren KG, Hechingen, und [b] Institut für Chemische Verfahrenstechnik, Universität Stuttgart, BRD

Today, blood purification by means of extracorporeal membrane-containing devices has become a state-of-the-art therapy for acute and chronic renal failure. The steady evolution of the membranes used in these processes for nearly 50 years has led to devices which are reliable, efficient and cost-effective.

The continuous development of the process as well as the membranes employed is illustrated in figure 1. The membrane used in the first hemodialyzer, which was successfully applied to humans in renal failure therapy, was a tube made from regenerated cellulose. With the introduction of hemofiltration, new membrane transport properties were required. These transport properties would best be accomplished with 'skin-type' asymmetrically structured membranes, which had been developed earlier for industrial application of ultrafiltration. These membranes were prepared as flat sheets or as hollow fibers from a variety of different synthetic polymers by a process referred to as phase inversion. Hemofiltration with asymmetric ultrafiltration membranes led to a significantly better clearance of high molecular weight toxins. Further development of renal failure therapy was achieved by combining hemodialysis and hemofiltration. This again required membranes with special structure-related transport properties. Such membranes were consequently developed from various polymers, such as polyacrylonitrile, polysulfone and polyamide.

During the last 10 years membranes based on polyamide have been adapted to the current clinical requirements. In the beginning of the 1980s,

[1] The authors are grateful to Rita Enderle and Roswitha Kunas for assistance in the preparation of the manuscript.

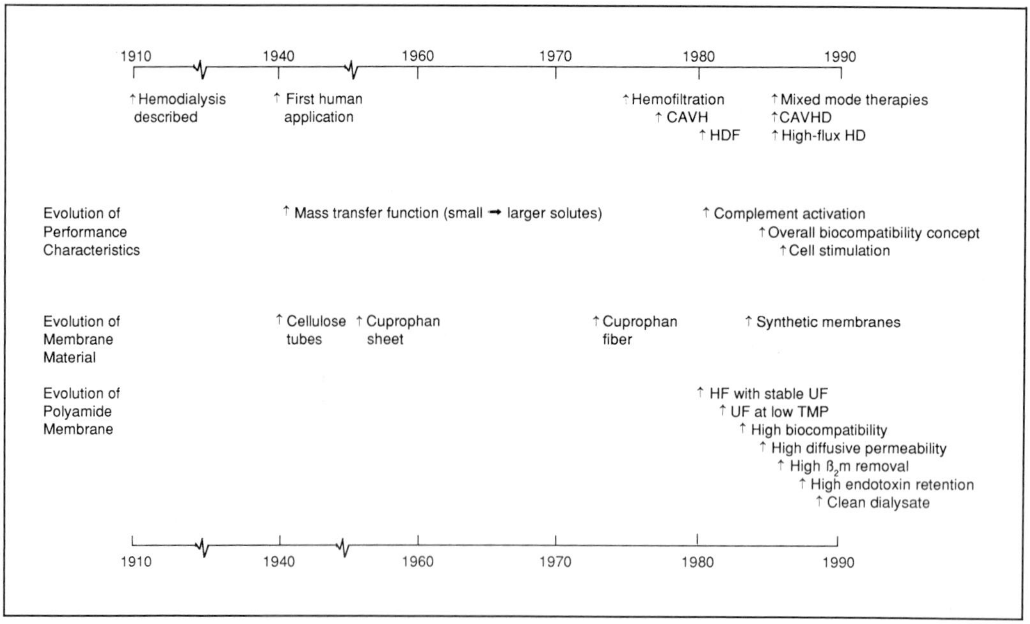

Fig. 1. Historical overview of dialysis treatment and membrane development. HD = Hemodialysis; HF = hemofiltration; HDF = hemodiafiltration; CAVH = continuous arteriovenous hemofiltration; CAVHD = continuous arteriovenous hemodiafiltration.

the polyamide membrane was developed as a high-permeable asymmetric membrane for hemofiltration processes (HF and CAVH), first as a flat sheet [1] and later as a hollow fiber membrane [2]. High filtration capacity was required to allow the exchange of large volumes of fluid in a reasonable time. Specifically for CAVH, but also for conventional hemofiltration, it is desirable to have a high filtration rate already at a low transmembrane pressure and to keep the high filtration rate constant during the entire treatment time. This was achieved by increasing the hydrophilicity of the polyamide membrane which reduced the protein-membrane interaction [3]. The original hydrophobic polyamide membrane was thus replaced by a more hydrophilic polyamide in all hemofilters. From the time that complement activation could be measured in hemodialysis, improved blood compatibility became a clinical request. This could be achieved by selection of a suitable polyamide polymer mixture. Further optimization of the hydrophilicity and the membrane structure resulted in a high diffusive permeability which allowed the use of the membrane also in high-flux dialysis [4]. Thus hemodiafilters, combining high diffusive and high hydraulic permeability, could be produced with this modified polyamide membrane. In the late 1980s the removal of higher molecular weight substances, such as β_2-microglobulin, became clinically desirable, and the membrane was tailored in terms of pore size to accomplish these requirements [5]. All polyamide membranes used for blood purification were optimized for β_2-microglobulin removal.

Recently, it has become apparent that a dialysis membrane should have the capability to reject bacterial products derived from contaminated dialysis fluid. Membranes of polyamide fulfill these requirements by combining high rejection/adsorption capacity for endotoxins with high permeability. They can therefore be used for high-flux dialyzers as well as for ultrafilters [6, 7].

In spite of the significant advances made during the last 40 years in both processes and membranes, further improvements are highly desirable. Compared to the human kidney all devices used to date are still lacking in performance, both as far as their efficiency of selective removal of toxic blood components as well as their blood compatibility is concerned. Polymer modification and membrane preparation techniques developed in recent years have the potential for further tailoring of the membrane transport properties and blood compatibility to be optimized for the different processes in renal failure therapy.

Nevertheless, the basis of all further development is a clear understanding of the blood-polymer interaction, of the relation between membrane structure and mass transport and of the mechanism of membrane structure formation. The structure- and surface-determined transport properties of membranes are again closely related to the processes in which they are used, thus process design and membrane development are also closely interrelated.

Fundamentals for Membrane Design

Membrane Processes in Renal Failure Therapy

In renal failure therapy membranes are used today in three different operating modes as indicated in the schematic flow diagrams of figure 2. In the first operating mode (a), referred to as hemodialysis, the dialysis process is carried out under isobaric conditions with a concentration gradient

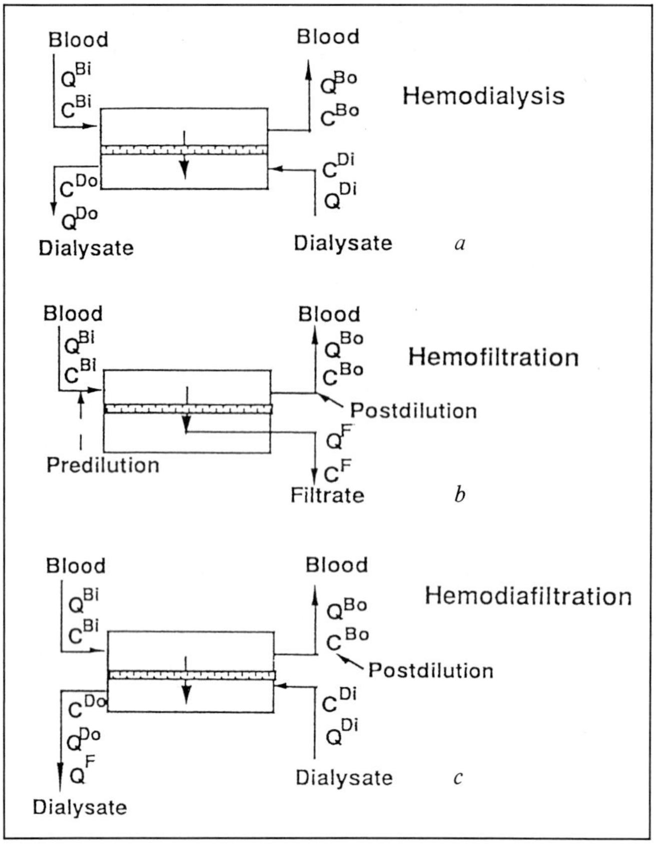

Fig. 2. Flow diagrams of different forms of renal therapy.

being the only driving force for the transport of solutes from the bloodstream into the dialysate. The second operating mode (b), referred to as hemofiltration, is a filtration process using a hydrostatic pressure difference as the driving force for the transport of fluid from the bloodstream. Solutes are transferred across the membrane by convection into the filtrate. This second mode of operation leads to a volume change of the bloodstream, which has to be replaced by a suitable solution either before (predilution), or after (postdilution) the filtration procedure. The third mode of operation (c) is referred to as hemodiafiltration and represents basically a combination of hemodialysis and hemofiltration. In clinical practise the membrane unit is generally regarded as a 'black box' and its solute removing capacity is expressed in terms of the 'clearance' C_L (ml/min).

$$\text{For hemodialysis: } C_L = \frac{(C^{Bi} - C^{Bo})Q^{Bi}}{C^{Bi}}. \tag{1}$$

$$\text{For hemofiltration: } C_L = \frac{C^F}{C^{Bi}} Q^F, \tag{2}$$

$$\text{with apparent sieving coefficient: } S = \frac{2C^F}{C^{Bi} + C^{Bo}}. \tag{3}$$

$$\text{For hemodiafiltration: } C_L = \frac{(C^{Bi} - C^{Bo})Q^{Bi} + Q^F C^{Bo}}{C^{Bi}}. \tag{4}$$

The overall performance of a blood purification device is determined mainly by the mass transport properties of the membrane which again are closely related to the membrane structure. To improve the properties of the membranes used in renal therapy an understanding of the function of the membranes in a mass separation process is essential.

Definition and Function of Synthetic Membranes in Mass Separation

Synthetic membranes may be rather different in terms of their structure, their basic material and their mass transport properties. Mass transport in synthetic membranes can be described by phenomenological equations which relate membrane structure properties to the mass transport. By introducing the proper relations for the driving forces applied in hemodialysis and hemofiltration, respectively, membrane mass transport can be adequately described [9] by:

$$J_i = \frac{\varepsilon r^2}{8\eta^\tau} \frac{\Delta P}{\Delta z} \frac{C_i^{MB} \exp\left\{\frac{\varepsilon r^2 \Delta P}{8\eta D_i^M}\right\} - C_i^{MD}}{\exp\left\{\frac{\varepsilon r^2 \Delta P}{8\eta D_i^M}\right\} - 1}, \tag{5}$$

where J_i is the flux of a component *i* through a membrane, ε the porosity of the membrane, *r* the radius of the membrane pores, η is the dynamic viscosity of the fluid in the membrane, τ is a tortuosity factor, ΔP a hydrostatic pressure difference between the two phases separated by the membrane, Δz is the thickness of the membrane, D_i^M is the diffusion coefficient and C_i^M the concentration of the component *i* in the membrane interphase. Thus, C_i^{MB} and C_i^{MD} are the concentrations of the component *i* in the membrane on the side facing the bloodstream and the dialysate, respectively.

Equation (5) describes the transport rate of a component *i* under conditions of combined hemodialysis and hemofiltration, i.e. hemodiafiltration, with the driving force consisting both of a hydrostatic pressure and concentration gradient. For the limiting case of vanishing hydrostatic pressure (equation 6), and vanishing concentration gradient in the membrane (equation 7), equation (5) reduces to:

$$\lim_{\Delta P - 0} J_i = \frac{\varepsilon D_i^M S}{\tau \Delta z} (C_i^B - C_i^D), \tag{6}$$

$$\lim_{\frac{dc}{dz}} \rightarrow OJ_i = \frac{\varepsilon r^2 S C_i^B}{8 \eta t} \frac{\Delta P}{\Delta z}. \tag{7}$$

Equations (6) and (7) describe, respectively, the transmembrane fluxes in hemodialysis and hemofiltration. From these equations it can be seen that: (1) In hemodialysis and hemofiltration the transmembrane transport rate is proportional to the porosity of the membrane and the sieving coefficient, and inversely proportional to the thickness of the membrane. (2) In hemodialysis the transmembrane transport rate is furthermore effected by the diffusion coefficient and the dialysate concentration. (3) In hemofiltration the transmembrane transport rate is effected by hydrostatic pressure differences across the membrane, the radius of the pores in the membrane and the viscosity of the fluid in the membrane pores. (4) In hemofiltration the transmembrane transport rate is independent of the molecular weight of the permeating components as long as their sieving coefficients are close to unity. (5) In hemodialysis the transmembrane transport rate is decreasing with increasing molecular weight of the permeating components, even if their sieving coefficients are one, because of decreasing diffusivity with increasing component size. The relation between particle size and its diffusivity is given by the Stokes-Einstein equation:

$$D_i = \frac{RT}{6\pi \eta r_i N_A}, \tag{8}$$

where D_i is the diffusion coefficient of a component *i* in water, R the gas constant, T the absolute temperature, π the constant, η the viscosity of the diffusion matrix, r_i the radius of the diffusing particle, and N_A Avogadro's number. The radius of a particle can be related to the molecular weight of a component.

So, in hemodialysis the transport rate of low-molecular-weight components across the membrane will always be higher than that of high-molecular-weight substances. However, in hemofiltration the transmembrane flow rate is independent of the molecular weight of the transported components for all solutes with a sieving coefficient of 1.

Membrane Preparation

Based on theoretical considerations of the mass transport in the membrane and the boundary layer, the structures of membranes with optimal performance in hemodialysis, hemofiltration and hemodiafiltration can be defined as follows:

(1) The active membrane layer should be as thin as possible to obtain high transmembrane fluxes. (2) The porosity of the membrane at the surface as well as in the matrix should be as high as possible to provide high transmembrane fluxes. (3) The diameter of the pores should not exceed a certain value to eliminate leakage of larger essential molecules. (4) The pore size distribution should be as narrow as possible to obtain a sharp molecular weight cut-off, i.e. sieving coefficients close to unity for all molecules up to a certain size. (5) The membrane structure should guarantee a certain minimum mechanical strength. (6) The diffusion coefficient in the membrane should be high. (7) The membrane protein adsorption should be limited. (8) The membrane should have good blood compatibility. (9) All materials of the final membrane have to be nontoxic and chemically inert.

An attractive way of preparing membranes with the above mentioned properties is employing a polymer precipitation process, generally referred to as phase inversion. Essentially all membranes used today in hemodialysis and hemofiltration are manufactured according to this principle.

Preparation of Phase Inversion Membranes

The membrane formation process consists of three consecutive steps. In the first step a polymer is dissolved in an appropriate solvent or solvent mixture to form a homogeneous, more or less viscous, solution containing 10–20 wt% polymer. In the second step the polymer solution is cast into a film. In the third step a nonsolvent is added, which leads to the conversion of the homogeneous, liquid polymer solution into a two-phase system with a solid, polymer-rich phase forming the rigid membrane structure and a liquid, polymer-poor phase forming the membrane pores. Phase separation can be achieved with any polymer mixture, which under certain con-

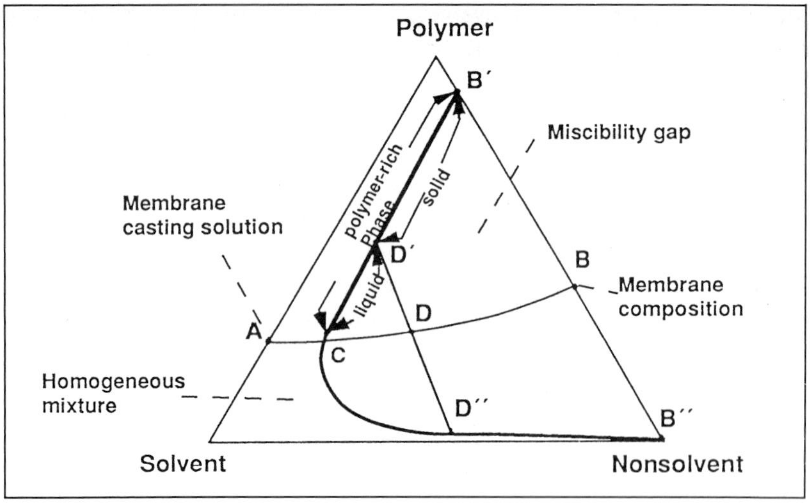

Fig. 3. Schematic ternary phase diagram showing the formation of a microporous membrane by addition of a nonsolvent to a homogeneous polymer solution.

ditions of temperature and composition forms a homogeneous solution, and then at a different composition separates into two phases. To obtain a microporous medium such as a membrane, however, requires both phases to be coherent. If the liquid phase is incoherent, a closed-cell foam structure will be obtained. If the solid phase is incoherent, a polymer powder will be obtained instead of a rigid structure. The membrane formation process is illustrated in figure 3, which shows the ternary phase diagram of a polymer, a solvent and a nonsolvent mixture. The corners of the triangle represent the pure components. Any point on the connecting lines between two corner points represents a mixture of two components and any point within the triangle represents the mixture of all three components.

The phase diagram of the three-component mixture shows a miscibility gap over a large range of compositions. If a nonsolvent is added to a homogeneous solution consisting of polymer and solvent, as indicated by point A on the solvent-polymer line, and if the solvent is removed from the mixture at about the same rate as the nonsolvent enters, the composition of the mixture will change following the line A–B. At point C, the composition of the system will reach the miscibility gap and two separate phases will begin to form, i.e. a polymer-rich phase represented by the upper

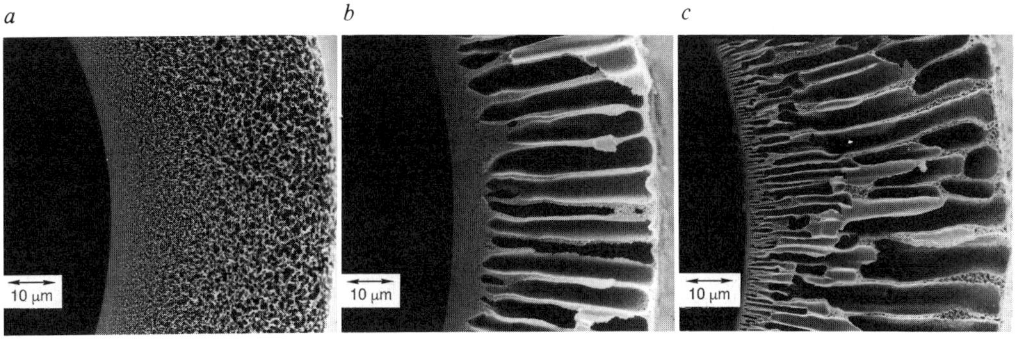

Fig. 4. Scanning electron micrographs of membrane cross-sections with typical structures. *a* Asymmetric membrane with 'sponge'-type structure, a dense skin at the surface and a porous substructure with increasing pore diameters from the top to the bottom side of the membrane. *b* Asymmetric membrane with a 3-layer structure, consisting of a dense skin at the surface, supported by a porous 'sponge'-type intermediate layer. *c* 'Finger'-type structure and a dense skin at the surface. Membrane sections were freeze-dried and coated with Au to minimize artefacts and shrinkage. Structure (*b*) represents the polyamide membrane used for high-flux hemodialysis, hemodiafiltration and hemofiltration. Structure (*c*) represents the polyamide membrane used for ultrafiltration.

boundary of the miscibility gap, and a polymer-poor phase represented by the lower boundary of the miscibility gap. At a certain composition of the three-component mixtures, the polymer concentration in the polymer-rich phase will be high enough to be considered as solid. This composition is represented by point D in figure 3. At this point the membrane structure is more or less determined. Further exchange of solvent and nonsolvent will lead to the final formation of the membrane, the porosity of which is determined by point B, which represents the mixture of the solid polymer-rich phase B′ and the liquid phase B″.

The phase separation in polymeric systems is determined by thermodynamic and kinetic parameters, such as the chemical potentials and diffusivities of the individual components and the Gibb's free energy of mixing of the entire system. Identification and description of the phase separation process is the key to understanding the membrane formation mechanism, a necessity for optimizing membrane properties and structures.

The different membrane structures obtained by phase inversion can be studied using scanning electron microscopy techniques. Figure 4 shows the cross-section of three different membranes prepared from a polyamide casting solution by the phase inversion process.

Table 1. Hydraulic and diffusive properties and technical data for the polyamide (PA) series of membranes

Membrane type	Membrane structure	Fiber inner diameter µm	Fiber effective length cm	Fiber wall thickness µm	Hydrophilicity	UF[c] water 37 °C ml/m^2/h/ mm Hg	UF[c] blood* 37 °C ml/m^2/h/ mm Hg	Pm urea 37 °C 10^{-4} cm/s
PA for high-flux hemodialysis and hemodiafiltration	3-layer	220	21	50	high	160	34	13
PA for hemofiltration	3-layer	215	25 (14 for CAVH)	60	high	160	34	–
PA for water ultrafiltration	skinned finger	215	25	60	low	400	–	–

* At TMP 50 mm Hg and γ_w (shear rate) = 2,278 s^{-1}.
– = Not important for these membranes types.

The mechanism of membrane phase formation by inversion is today well understood. By choosing the proper preparation parameters, the various structures can be made from almost any polymer. This means that, in terms of the basic technology, it is possible to meet all the structural requirements of an ideal membrane for renal therapy [8, 9]. As far as biocompatibility is concerned the selection of the membrane polymer is of prime importance, since this determines to a large extent the interaction between the blood and the membrane surface.

In recent years polymer blends became a convincing concept for tailor-made membranes. This concept facilitates a balancing of beneficial membrane structure and surface properties. For the preparation of membranes to be used in blood detoxification processes polymer blends consisting of a mixture of hydrophobic and hydrophilic components are used.

The blending/alloying concept has allowed the preparation of problem-oriented membranes in a series of polyamide-based structures. Currently, there are three different types of these membranes manufactured by Gambro which are used in renal failure therapy: (1) a high-flux dialysis membrane; (2) a hemofiltration membrane, and (3) an ultrafiltration membrane to be used for the preparation of infusion solutions and dialysate.

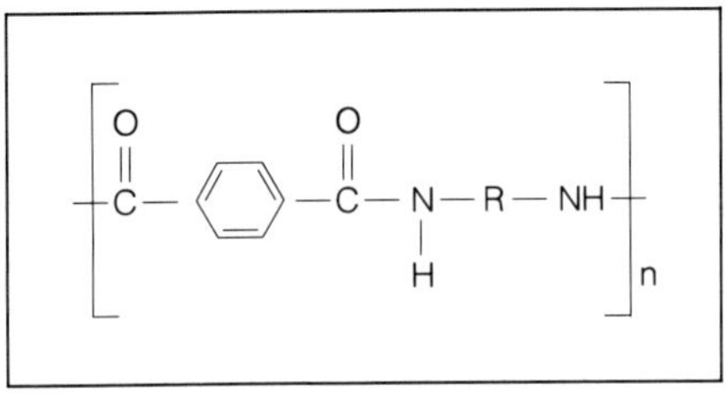

Fig. 5. Chemical structure of aliphatic/aromatic polyamide. R = 1:1 isomere mixture of 2,2,4- and 2,4,4-trimethylhexamethylenediamine.

The properties of the three different membranes, which are all prepared from the same basic polymer are summarized in table 1. The structure is the same for the first and second membrane types (fig. 4b). The structure of the third membrane is shown in figure 4c.

Selection of Polymers for the Preparation of Membranes for Renal Failure Therapy

For all medical membrane applications the selection of the membrane material is of prime importance. The polymer or polymer mixture must not only meet the requirements for preparing tailor-made membrane structures but must also have good blood compatibility at the surface in the final configuration. An aromatic-aliphatic copolyamide (fig. 5) was chosen as the hydrophobic component in the polyamide membranes listed in table 1.

This polymer provides good mechanical stability and film-forming properties. It is amorphous in structure and contains a minimum of monomers or low-molecular-weight components that could be leached into the bloodstream during membrane application [10]. In contrast to other polyamides, particularly aliphatic ones, this copolyamide exhibits excellent blood compatibility with a minimum of complement activation, protein adsorption and blood clotting. It can also be blended easily with polyvinylpyrrolidone (PVP), a hydrophilic water-soluble polymer, which is used as the second component in the PA membranes. PVP also has excellent biocompatibility reflected by intravenous application as a plasma expander in large amounts without any adverse effects [11]. Hydrogen bonds between the PVP and the polyamide provides good miscibility of the two polymers over a large concentration range, and consequently during the actual membrane precipitation step small domains, in the range of a few 100 nm, of hydrophilic and hydrophobic polymers are formed. Because of strong hydrogen bonds the PVP is not eluable out of the membrane during normal hemodialysis or hemofiltration conditions. The third component needed in the membrane preparation process is the solvent to prepare the casting solution. A solvent which is particularly suited for the preparation of the casting solution for the PA-series membranes is dimethylsulfoxide (DMSO). It is not only a good solvent for the polymer blend, but it is also easily removed due to its high water solubility from the final membrane below detection level. Its biological toxicity is reported to be very low [12].

The selection of the polymer blend composition, its concentration in the casting solution, the composition of the precipitation fluid and the precipitation temperature determine the structure and thus the properties of the final membrane. The proper selection of these parameters leads to membrane structures consisting of three clearly distinguishable layers as indicated in the scanning electron micrograph of figure 6. This shows the cross-section in different magnifications of this polyamide hollow fiber membrane for hemodialysis and hemofiltration.

Characterization of the Polyamide Membranes

The performance of a membrane in renal failure therapy is determined by its structure, its overall mass transfer properties, and its blood compatibility.

Membrane Structure Characterization

The polyamide membrane has an integral 3-layer structure (fig. 6). At the blood contacting surface it has a 0.1–0.5 μm thick skin with a very high surface porosity. This skin also provides a narrow pore size distribution which basically mediates the steep decline of the sieving coefficients as illustrated in figure 8. This selective layer is supported by a 5 μm thick sponge-type structure with the same high porosity, but with increasing pore diameters from the surface towards the bulk structure. The sponge structure is supported by a finger-type structure. During the application this finger-structure is filled with water and provides almost no hydrodynamic resistance. This promotes high transmembrane fluxes and also high diffusive permeability, close to that in aqueous solution.

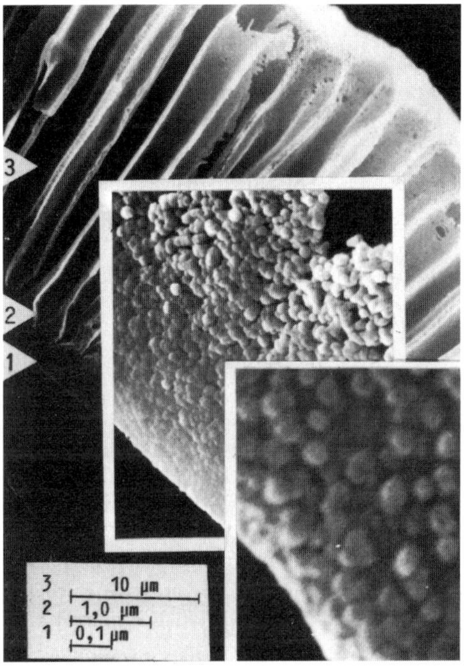

Fig. 6. Scanning electron micrograph of a cross-section of the 3-layer polyamide membrane used for hemodialysis, hemodiafiltration and hemofiltration.

The inner surface of the fiber, which is in contact with blood is very smooth even under high magnification. Sieving coefficient measurement with different substances like dextrans and proteins give a notion that the mean pore diameter of the polyamide membrane in the skin layer is in the range of 5 nm. Since the smallest diameter of the pores is on the inner surface, the penetration of proteins into the membrane is prevented. Thus, the membrane shows very little decrease in permeability or reduction in sieving coefficient during the treatment time.

Hollow Fiber Membrane Dimensions

The mass transport properties of a hemofilter or a dialyzer are not only determined by the membrane properties, but, due to flowdynamic considerations, they are also influenced by the dimensions of the hollow fibers. To achieve optimal reduction of the boundary layer on the blood

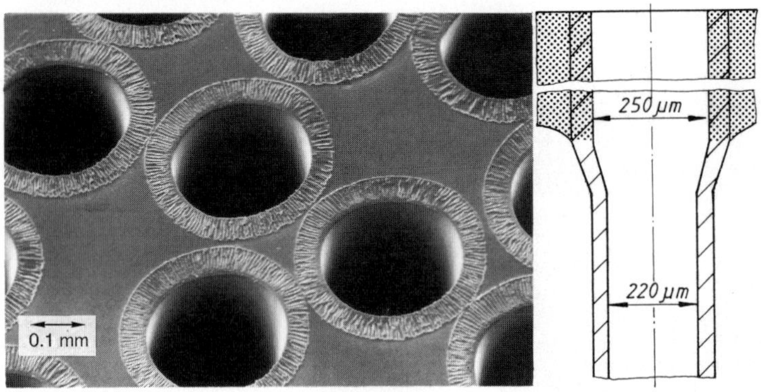

Fig. 7. To the left a scanning electron micrograph of potted polyamide hollow fibers at the blood inlet surface. To the right a schematic cross-section of a single potted hollow fiber (see text).

side and optimal shear rate at an optimal pressure drop, the fiber inner diameter and the length of polyamide hemofilter [14] and dialyzers are optimized as documented in table 1. An inner diameter of 220 µm was chosen for polyamide dialyzers for optimal clearance and pressure drop, based on a mean hematocrit of 30% [13].

According to the Hagen-Poiseuille law the pressure drop ΔP is inversely proportional to the fourth power of the fiber inner diameter d^4. This means that a small increase of the inner diameter gives a large reduction of the pressure drop. Reducing the pressure drop on the blood side of devices used in hemodialysis is desirable to avoid backfiltration. In a hollow fiber with a diameter of 220 µm the pressure drop on the blood side is 46% lower than in a 200 µm fiber. The minimum filtration, which is needed to avoid backfiltration, is about 20 ml/min lower in a 220-µm inner diameter fiber than in a comparable 200-µm inner diameter fiber assuming the same surface area and permeability.

Good rejection properties of the polyamide series of membranes for endotoxins and bacteria reduce the risk of contamination from the dialysate. About 90% of the transport of low-molecular-weight endotoxin ($<$ 1,000 Da) through the membrane is by diffusion, whether a low-flux or a high-flux membrane is used [15]. Since polyamide retains low molecular substances by adsorption, the risk of contamination is significantly lower when polyamide high-flux membranes are used instead of other dialysis membranes [6, 16].

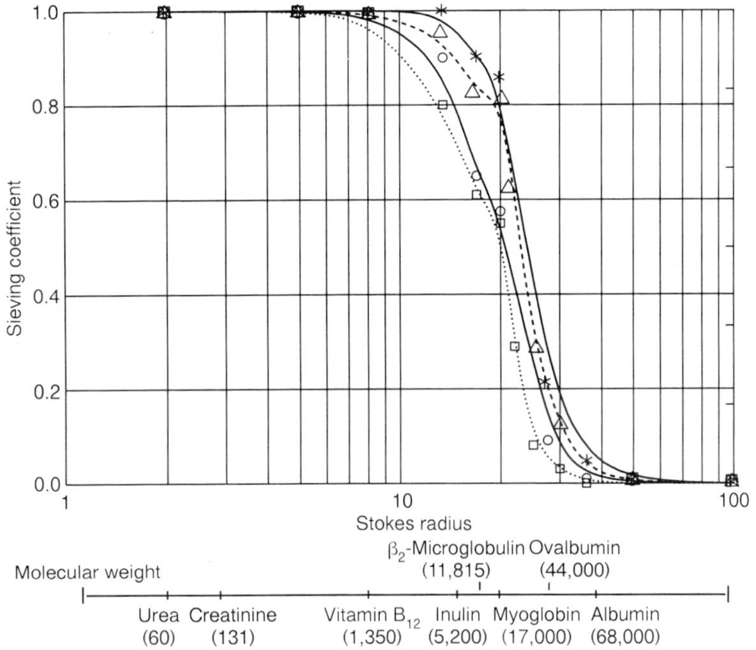

Fig. 8. Sieving coefficients of the hydrophilic polyamide membrane determined with dextran and protein model solution before and after contact with plasma. △ = Dextran before plasma; □ = dextran after plasma; * = proteins before plasma; ○ = proteins after plasma; concentrations: 0.01%, plasma flow Q^b = 300 ml/min; Q^F = 60 ml/min; membrane surface area 1.3 m^2 , γ_w = 2,278 s^{-1}.

For their application in renal failure therapy, the hollow fiber membranes are installed in a housing. This is usually done by potting the fibers at the end of the tubular outer shell using appropriate potting material, such as polyurethane. The very open finger-like structure of the polyamide membranes allows the potting material to penetrate the pore structure of the hollow fibers up to their inner skin. This leads to a very smooth surface of the blood inlet and outlet areas, as shown in figure 7. The combination of the potting material and the fiber structure furthermore leads to larger fiber inlet diameter, about 250 μm, compared to the actual fiber diameter, about 215 and 220 μm, respectively. Figure 7 schematically illustrates a cross-section of a polyamide filter potted with polyurethane. Both blood inlet design and surface smoothness beneficially affect the thrombogenicity.

Table 2. Contact angles of the hollow fiber membranes

Membrane	Advancing contact angle	Receding contact angle
Polyamide for blood treatment	67	38
Polyamide for water filtration	78	59

Membrane Hydrophilicity and Protein Interaction

To minimize the interaction of proteins with the membrane surface is desirable for several reasons. Proteins adsorbed at the membrane surface affect the performance of the membrane and may lead to poor blood compatibility. The molecular structure of the blood proteins contains different surface domains which can interact with the membrane surface by hydrogen bonds, ionic and hydrophobic forces [17]. Therefore, protein adsorption on the membrane surface is very strongly influenced by the hydrophilic/hydrophobic character of the membrane surface.

Studies performed by Elwing et al. [18] relate the degree of hydrophilicity of a surface to its protein adsorption behavior. Monitoring ellipsometrically the amount of adsorbed human IgG on a continuous hydrophilicity gradient, a sharply reduced IgG adsorption was found when the contact angle of a water droplet at the dry membrane surface was between 40 and 70°. Contact angle measurements as a tool to determine the hydrophilicity of a surface are described elsewhere [19]. They can also be applied to characterize membrane surfaces in terms of their wettability.

On a heterogenous surface, as in a polyamide/PVP membrane, the contact angle is different for the advancing and the receding interface. The difference (hysteresis) is a function of the dispersion of hydrophilic and hydrophobic sites at the surface. Table 2 shows that the polyamide membranes for hemodialysis or hemofiltration exhibit a contact angle in the range where protein adsorption and cell interaction were reported to be minimal [19]. On the other hand, the membrane used for water filtration has been tailored to lower hydrophilicity and is optimized towards higher ultrafiltration and rejection rates for endotoxins. The contact angle is therefore larger.

Membrane Surface Charges

Surface charges may also affect the performance and the blood compatibility of membranes. Based on the molecular structure of the polyamide and from ZETA-potential measurements, it can be concluded that the polyamide series of membranes are almost neutral. The ZETA potential of a polyamide flat sheet membrane, determined after an 8-hour wetting test at a pH value of 7, is approximately 7 mV, compared to 15 mV obtained for a cuprophan membrane under the same test conditions.

Sieving Properties

The sieving coefficient, as defined according to equation 3, is determined mainly by the pore size and the pore size distribution of the membrane. It is additionally influenced by the interaction of proteins with the membrane surface and the membrane structure.

To determine the effect of membrane-protein interaction on membrane sieving, the sieving coefficient of a membrane before and after contact with plasma was determined with dextran and protein model solutions. The test results are shown in figure 8, where the sieving coefficient is plotted versus the Stokes radius of the respective model solution component.

These test results indicate that the sieving coefficients determined for dextran and protein model solutions and polyamide membranes before and after plasma contact do not differ to a major extent. Thus, protein adsorption to these membranes has only a minor effect on the sieving of the membrane. In addition, sieving coefficients measured in vitro in plasma correlate well with sieving coefficients measured in vivo, e.g. for β_2-microglobulin [20]. In clinical application the sieving coefficients of the polyamide membranes were found to be nearly constant during the entire treatment period [5, 21]. This is an impressive proof of the membrane concept. The structure of the polyamide membrane is truly asymmetric, with the pores having the smallest diameter at the surface of the membrane and thus preventing proteins from entering and plugging the membrane structure. For other membranes changes of the sieving coefficient, especially for high molecular weight substances, are documented in in vivo application [22]. The sieving properties of the polyamide membranes are optimized to provide a mean β_2-microglobulin sieving coefficient of 0.6 during the duration of clinical treatments. This guarantees a high β_2-microglobulin removal, especially in hemofiltration and hemodiafiltration, where the large amount of ultrafiltrate gives a corresponding large convective transport [22]. In spite of the high removal rate of β_2-microglobulin, the total protein loss is remarkably low. The polyamide membranes have a sieving coefficient for albumin (MW 68,000 Da) of below 0.001. This assures a total protein loss of less than 0.05 g/l filtrate during a hemofiltration procedure [24].

Performance in Hemofiltration

The filtration rate of a hollow fiber hemofilter for whole blood depends on the composition of the blood (blood cell and protein concentration), the blood flow velocity in the fiber lumen and on the transmembrane pressure (TMP). At low TMP there is a linear increase of the filtration rate with increasing TMP. Further increase of the pressure beyond a certain value leads to excessive concentration polarization and thus to the formation of a protein layer at the membrane surface, which affects the filtration rate. At a certain TMP the transmembrane flux will no longer increase with a pressure increase [25].

The high surface porosity, the hydrophilicity and the low skin thickness of the PA membranes lead to high constant filtration rates at low TMP. These properties are of great significance for HF and CAVH. At a given blood composition and constant TMP the filtration rate can be increased by increasing the wall shear rate by means of higher blood flow velocities in the hollow fiber membrane.

To achieve maximum performance of hemofilters in terms of filtration rates and thus clearance, the filter must be optimized in terms of the membrane area per filter, the inner fiber diameter and the fiber length. The blood flow velocity must be adjusted to the fiber length and diameter. The effect of fiber and filter dimensions and the blood flow rate on the filtration rate is illustrated in figure 9, where the filtration rate of a polyamide hemofilter operated at three different blood flow velocities is plotted versus the transmembrane pressure.

Diffusive Permeability

Equations 5 and 6 give the basic principles for the polyamide membrane design: high membrane porosity, short diffusive distance and high diffusion coefficient in the membrane. These factors yield a high diffusive permeability and high clearance values over a wide range of molecular weight.

The diffusion coefficient D_{H_2O} of molecular components in water depends on the Stokes radius of the component, as indicated in equation 8. The diffusivity through a membrane is reduced, due to an additional mechanical resistance, and is expressed as effective diffusion coefficient D_{Meff}. The membrane diffusive permeability P_m is related to membrane

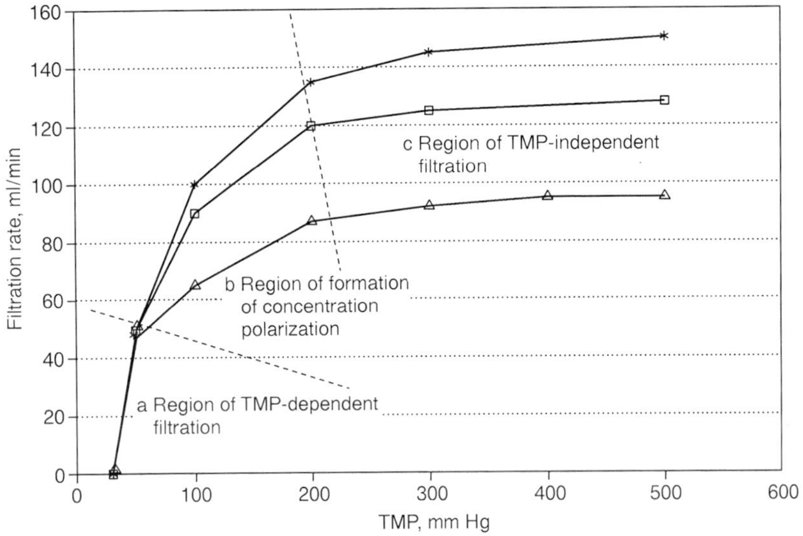

Fig. 9. Filtration rate UF of PA hemofilter (hydrophilic) as a function of the TMP at different blood flow and shear rate. Hematocrit = 28%; protein concentration = 6 g/l. The inner diameter of the hollow fiber is 215 μm, its length 25 cm and the total area of the filter is 1.4 m^2 (FH77H). –∗– Q^B = 200 ml/min, γ_w = 1,520 s^{-1}; –□– Q^B = 300 ml/min, γ_w = 2,278 s^{-1}; –△– Q^B = 400 ml/min, γ_w = 3,040 s^{-1}.

diffusivity by $P_m = D_{Meff}/\Delta z$, where Δz is diffusion distance. It is the reciprocal of the membrane resistances $R_M = 1/P_m$.

The total resistance R for the diffusive transport in a hemodialyzer is the sum of the resistance of the actual membrane and the resistances in the laminar boundary layers on both membrane sides. $R = R_M + R_B + R_D$, where R is the diffusive resistance of the dialyzer, R_M that of the actual membrane and R_B and R_D, respectively, are those of the boundary layers at the membrane surfaces facing the bloodstream and the dialysate.

The resistance R_D is mainly affected by the dialysate flow distribution, i.e. by the fiber packing and the dialysate flow velocity. The determining resistance, however, is that of the membrane, assuming optimal packing. In PA membranes having the 3-layer structure the coefficients D_{Meff} are high and close to the diffusivities in pure water. Compared to low-flux membranes high diffusive coefficients are obtained over a large molecular weight range, because of more uniform pore sizes distribution and high overall porosity. This is illustrated in figure 10, which shows the diffusion coefficient of

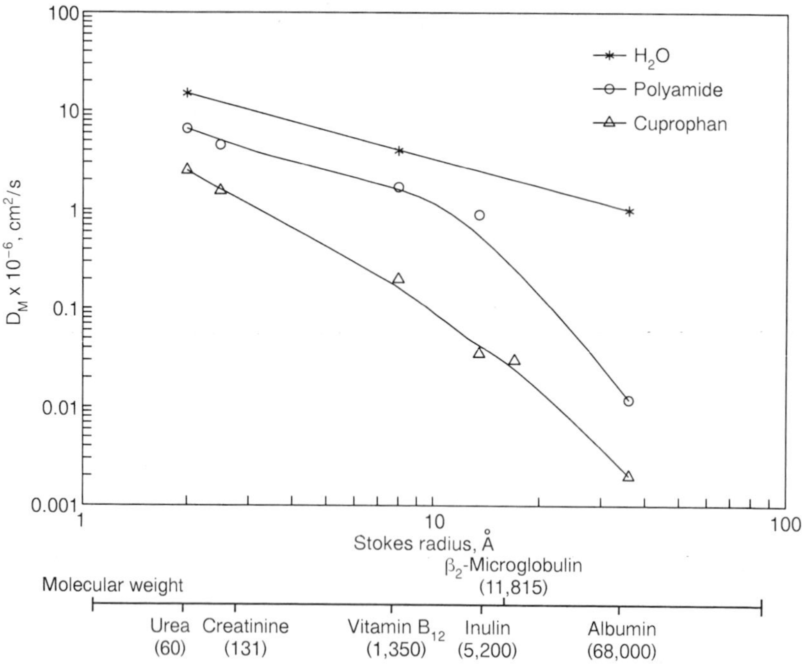

Fig. 10. Diffusion coefficient in water and effective membrane diffusion coefficient (D_M) for different size solutes. Solute molecular weights are in parentheses; D_M calculated with Δz for polyamide = 50 μm, for cuprophan = 16 μm (wet).

various molecular weight components in water and two different membranes as a function of both their molecular weight and Stokes radius.

The combination of high diffusive permeability and high convective permeability guarantees efficient treatment in the hemodialysis, as well as the hemodiafiltration and hemofiltration modes. High clearance values are achieved for low as well as for high-molecular-weight substances, including phosphate and β_2-microglobulin [4, 5]. This is illustrated in figure 11, which shows clearances measured with a cuprophan and a polyamide (Polyflux) dialyzer, both with 1.3 m^2 membrane area. The clearances were determined for pure diffusive permeation ($Q^F = 0$) and with additional filtration (Q^F = 60 ml/min). The filtration flux gives an additional contribution to the clearance, especially in the high molecular weight range. Due to the membrane structure and hydrophilicity the clearances are constant during the treatment time (fig. 12).

11

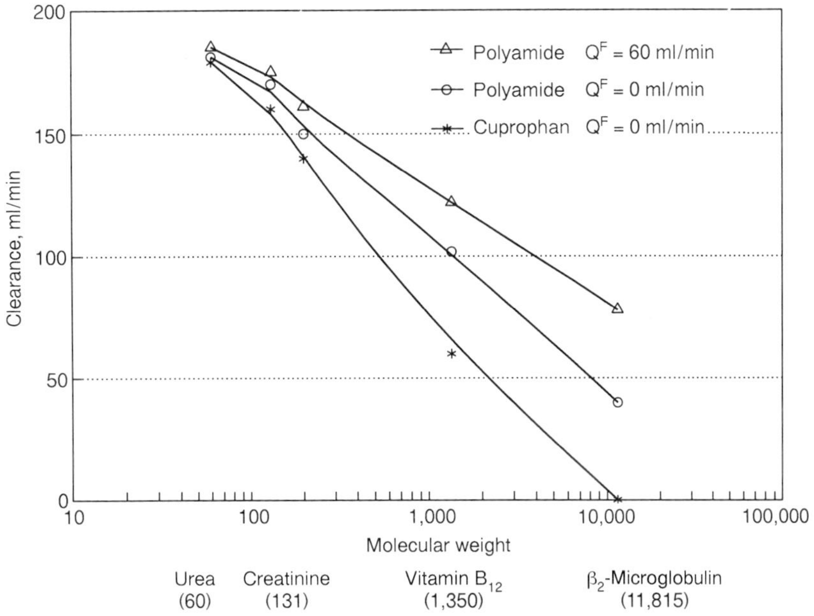

12

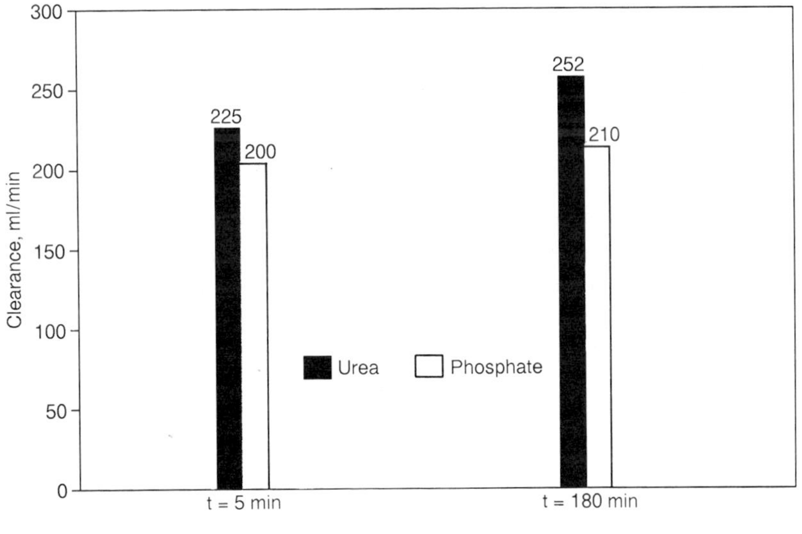

Fig. 11. Clearance of hydrophilic polyamide dialyzer (Polyflux 130) compared to a cuprophan dialyzer of the same surface area (1.3 m^2). Q^B = 200 ml/min; Q^D = 500 ml/min.

Fig. 12. In vivo clearance for urea and phosphate of polyamide hemodiafilter (Polyflux 130) at the beginning and the end of treatment. Q^B = 300 and 500 ml/min; Q^F = 20 ml/min. n = 8 patients. Data from Fournier and Morinière [4].

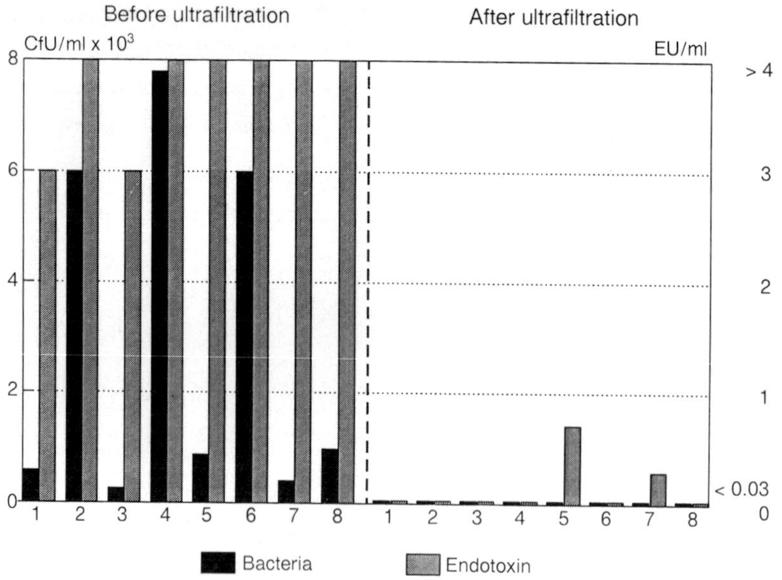

Fig. 13. Mean concentration of bacteria and endotoxin in dialysis fluid measured before and after ultrafiltration with a polyamide ultrafilter in 8 dialysis treatments. CfU = Colony-forming units; EU = endotoxin unit. Data from Bambauer et al. [27].

Ultrafilter for the Preparation of Dialysate and Infusion Solutions

In recent years it has become apparent that bacteriological contamination of the dialysate could lead to adverse patient reactions as suggested by the interleukin-1 hypothesis [28]. It was also found that the infusion solution in hemofiltration could contain low concentrations of endotoxins. Most commercially available ultrafiltration membranes do not meet the requirements in terms of endotoxin rejection for dialysate and infusion solutions. Thus, an optimized polyamide membrane was developed for this special application [26]. Two characteristic properties of the polyamide membrane are of importance: (a) the high water filtration rate, and (b) the ability to reject bacteriological material. The membrane morphology was especially adapted towards this application, by reducing the thickness of the inner membrane layer (as shown in fig. 4c) and by retaining the hydrophobicity of the membrane. The ultrafiltration rate for the polyamide ultrafilter at 20 °C is in the range of 400 ml/m²/h/mm Hg, as shown in table 1. The rejection of bacteria and endotoxin is given by the logarithmic reduction value of feed to permeate, which yields in the polyamide filter $> 10^7$ for bacteria and $> 10^4$ for viruses and endotoxins [7, 12, 26]. An example of the clinical application of this polyamide membrane for the removal of bacteria and endotoxins is shown in figure 13. Although the sieving cut-off of the membrane with water solution is in the range of about 40,000 Da, the membrane has the ability to reject endotoxins even in molecular weight range below 2,000 Da. The mechanism responsible for the removal of the low-molecular-weight endotoxins by polyamide filters is not yet fully understood. It can be speculated that adsorption is mainly caused by hydrophobic/hydrophilic interaction or by the high electron density in the aromatic ring of the polyamide or by hydrogen bridging of pseudomonas peptide-type exotoxins to the amide group.

Summary and Conclusion

Progress in membrane and membrane process development have contributed to the continuous improvement of the extracorporeal treatment of renal failure during the last 40 years. Today membranes can be adapted to the specific needs required by their clinical application.

Synthetic structures like the polyamide membranes offer a wide range of possibilities in performance and hemocompatibility, due to: (1) the use of block polymers or polymer blends and mixtures, and (2) modification of membrane structure, pore size and porosity, due to changes of the preparation process.

However, further improvements are still required to achieve the ultimate goal of matching the performance of biological membranes.

References

1 Gullberg CA, Göhl H, Bergkvist G, von Herrath D, Schaefer K: A new hemofiltration system. Proc Eur Soc Artif Organs 1977;IV:56–69.

2 Göhl H, Konstantin P, Gullberg CA: Hemofiltration Membranes. Contr Nephrol. Basel, Karger, 1982, vol 32, pp 20–30.

3 Olbricht CJ, Häbel U, Frei U, Koch KM: High performance CAVH without pumps, suction and redilution. Blood Purif 1986;4:222–223.

4 Fournier A, Morinière P: A new capillary hemodiafilter based on a Polyamide membrane: Polyflux 130. Clinical report Lund, Gambro, 1989, HCE 7497.

5 Flöge J, Granolleras C, Deschodt G, Heck M, Baudin G, Branger B, Tournier O, Reinhard B, Eisenbach GM, Smeby LC, Koch KM, Shaldon S: Highflux synthetic versus cellulosic membranes for β_2-microglobulin removal during hemodialysis, hemodiafiltration and hemofiltration. Nephrol Dial Transplant 1989;4:653–657.

6 Lonnemann G, Flöge J, Schindler R, Behme T, Lenzner B, Shaldon S, Koch KM: Passage of cytokine-inducing E-coli fragments (CEF) through various hemodialysis (HD) membranes (abstract). Kidney Int 1989;35:254.

7 Dinarello CA, Lonnemann G, Maxwell R, Shaldon S: Ultrafiltration to reject human interleukin-1-inducing substances derived from bacterial cultures. J Clin Microbiol 1987;25:1233–1238.

8 Strathmann H: Production of microporous media by phase inversion processes, in Lloyd DR (ed): Material science of synthetic membranes. ACS Symp Ser 269, Washington, 1985, pp 165–195.

9 Strathmann H, Göhl H: Membranes for blood purification – state of the art and new developments. Contr Nephrol. 1990. Basel, Karger, vol 78, pp 119–141.

10 Herold J, Meyerhoff G: Molecular weights and molecular weight distribution of poly (trimethylhexamethyleneterephthalamide) (Trogamid T), Makromol Chem 1980; 181:2625–2636.

11 Robinson BV, Sullivan FM, Borzelleca JF, Schwartz SL: A critical review of the kinetics and toxicology of polyvinylpyrrolidone (Povidone). Chelsea, Lewis Publishers, 1990, pp 147–149.

12 Martin D, Hauthal HG: Dimethylsulfoxid. Berlin, Akademie-Verlag, 1971, pp 453–463.

13 Sakai K: Technical determination of optimal dimensions of hollow fiber membranes for clinical dialysis. Nephrol Dial Transplant 1989;4(suppl):73–77.

14 Schneider H, Streicher E, von Mylius U: A theoretical and experimental approach towards optimal dimensions for capillary hemofilters. Artif Organs 1979;4(suppl): 114–118.

15 Colton CK: Analysis of membrane processes for blood purification. Blood Purif 1987;5:202–251.

16 Smeby LC, Göhl H, Nystrand R, Deppisch R: Diffusion and convection of endotoxins through low and high flux dialysis membranes. Submitted.

17 Andrade JD (ed): Surface and Interfacial Aspects of Biomedical Polymers. New York, Plenum Press, 1985, vol 2, pp 1–80.

18 Elwing H, Alkenda A, Ivarsson B, Nilsson U, Welin S, Lundström I: Protein adsorption on solid surfaces: physical studies and biological model reactions; in Brash HL, Horbet TA (eds): Proteins at Interfaces. Physiochemical and Biochemical Studies. Washington, ACS, 1987, Ser 343, pp 468–488.

19 Stier E, Deppisch R, Nawroth PP, Bierhaus A, Hänsch GM, Ritz E: Release of mediators and activation of protooncogenes in mononuclear cells is dependent on the physicochemical structures of dialysis membranes (abstract). J Am Soc Nephrol 1990;1:379.

20 Flöge J, Göhl H, Schindler R, Granolleras C, Deschodt G, Lonnemann G, Koch KM, Shaldon S: In vitro versus in vivo, evaluation of membrane β_2-microglobulin permeability. Blood Purif 1988;6:345–346.

21 Heinzler P, Schauseil M, Kutkuhn B, Chlebowski H: Grabensee B: Elimination of β_2-microglobulin by a new polyamide hemodiafilter. Abstr Book. ESAO 1988, p 79.

22 Röckel A, Hertel J, Fiegel P, Abdelhamid S, Panitz N, Walb D: Permeability and secondary membrane formation of a high flux polysulfone hemofilter. Kidney Int 1986;30:429–432.
23 Schäfer K, Kaiser JP, Von Herrath D, Göhl H, Hagemann J: The striking effect of a new polyamid hemofilter on the serum beta-2-microglobulin in hemodialysis patients (abstract). Kidney Int 1987;33:237.
24 Birk HW, Kistner A, Schütterle G, Wizemann V: Characterization of protein permeability of membranes used for hemofiltration and hemodialysis. Submitted.
25 Henderson LW, Colton CK, Ford CA: Kinetics of hemodiafiltration, II. Clinical characterization of a new blood cleansing modality. J Lab Clin Med 1975;85:372–391.
26 Nystrand R, Johansson NO: Rejection capacity of microorganisms and endotoxins in an ultrafilter (abstract). Blood Purif 1984;2:222.
27 Bambauer R, Jutzler GA, Schmidt R, Weber U, El-Saadi R, Oest A, Jung WK: Ultrafiltration of dialysis fluid for hemodialysis. Trans Am Soc Artif Organs 1989; 35:516–518.
28 Shaldon S, Deschodt G, Branger B, Granolleras C, Baldamus CA, Koch KM, Lysaght MJ, Dinarello CA: Hemodialysis hypotension: The interleukin hypothesis restarted. Proc Eur Dial Transplant Assoc 1985;22:229–243.
29 Mion C, Canaud B, Garred LJ, Stec F, Ngyen QV: Sterile and pyrogen-free bicarbonate dialysate: A necessity for hemodialysis today. Adv Nephrol 1990;19:275–314.

Hermann Göhl, Gambro Dialysatoren GmbH & Co. KG,
D-W-7450 Hechingen (FRG)

Shaldon S, Koch KM (eds): Polyamide – The Evolution of a Synthetic Membrane for Renal Therapy. Contrib Nephrol. Basel, Karger, 1992, vol 96, pp 26–46

Biocompatibility of the Polyamide Membranes

Reinhold Deppisch, Margarita Betz, Gertrud Maria Hänsch, Ernst-Wilhard Rauterberg, Eberhard Ritz

Departments of Immunology and Internal Medicine, University of Heidelberg, FRG

Biocompatibility Considerations in the Design of the Polyamide Membrane

Endothelial cells are the ideal nonthrombogenic and nonactivating surfaces for blood contact. For obvious reasons synthetic membranes cannot completely imitate what nature has achieved during evolution over millions of years; especially the dynamic regulation of surface properties of endothelial cells cannot be provided by artificial membrane surfaces. Cellulosic membranes were the first to be used for hemodialysis. Considerable progress has meanwhile been made in designing alternative membranes which are more biocompatible than the original cellulosic membranes. Nevertheless, many problems remain.

The development of the polyamide membrane dates back to the year 1976 when resilient membranes were needed for hemofiltration purposes, i.e. membranes capable of mechanically withstanding the high filtration pressures needed to give high ultrafiltration flow rates. A totally asymmetric 'skin' structure had initially been chosen with polyamide in homogenous isotropic orientation contacting the blood interface. This 'parent' membrane had remarkably good biocompatibility characteristics as documented by the low rate of production of the complement split products C3d, C3a and C5a [1–3], only minor release of granulocyte marker enzymes [1] and stable blood cell counts in the course of the treatment.

One disadvantage became apparent, however, soon after the introduction of the parent polyamide membrane. The sieving coefficient for large molecules, of which β_2-microglobulin (β_2m) was considered a marker, was relatively low and not constant over time [4, 5]. Using the parent hydrophobic polyamide membrane, Flöge et al. [6] noted an increase of the sieving coefficient for β_2m from 0 to a saturation level of 0.4 in the early phase of treatment. This reflects the change of membrane transport properties as a result of specific interactions of proteins with the surface.

When the connection between β_2m accumulation and dialysis-related amyloidosis was established, optimizing β_2m removal became a goal for further membrane development. Since the membrane characteristics which determine β_2m removal are also important determinants for biocompatibility properties, preserving or even enhancing the good biocompatibility of this parent polyamide membrane became a second objective.

In an effort to raise membrane hydrophilicity without risking deterioration of biocompatibility, a microdomain structure was chosen for the modified polyamide membrane. This new aliphatic-aromatic polyamide membrane is based on an alloy system between polyamide (PA) and polyvinylpyrrolidone (PVP). Hydrophilicity is conferred by the latter component [7]. As first pointed out by Okano et al. [8, 9], a microdomain structure confers an optimal 'trade-off' between the undesirable properties of hydrophobic surfaces, i.e. adsorption of proteins and cells, and the undesirable properties of hydrophilic surfaces, i.e. interaction with blood cells and plasma proteins causing activation or denaturation. In the polyamide membrane, hydrophobic and hydrophilic patches of approximately 100–500 nm across are closely intermingled [10, 11]. If the size of a hydrophobic island falls below a critical dimension, hydrophobic interactions with plasma components diminish sharply. This is readily explained by thermodynamic considerations. Stable hydrophobic interactions between the membrane surface and plasma constituents can only be established if the interaction is driven by a critical amount of free energy. If the size of a hydrophobic patch does not provide the threshold amount of activating free energy, hydrophobic interaction will no longer be possible. The same considerations apply for hydrophilic interaction. It is our current understanding that cells interact with hydrophilic membranes by establishing contact with and by penetrating into the uppermost nanometers of the membrane. The same holds true for plasma proteins. This interaction is thought to underlie activation of cells and denaturation of proteins with

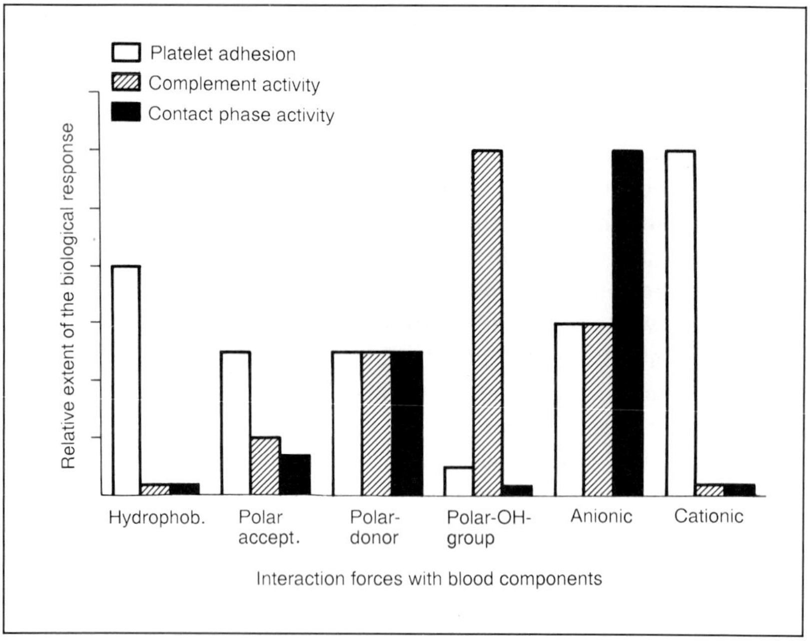

Fig. 1. General features of blood-surface interactions considering contact activation, complement activation and platelet adhesion as a function of physicochemical surface properties adapted from Matsuda [15].

or without the functional consequences of activation or inactivation [12–14].

When one constructs a hydrophilic/hydrophobic membrane with microdomain structure, some basic considerations apply as discussed in depth by Matsuda [15]. Figure 1 illustrates that hydrophobic, polar and ionic structures have antagonistic actions on the contact phase of coagulation, complement activation and platelet adhesion, respectively. On the one hand, hydrophobic membranes do not activate the contact phase of coagulation and complement, but cause strong platelet adhesion. On the other hand, hydroxyl groups which primarily interact by hydrogen bonding while displaying favorable adhesion or nucleophilic properties, are strongly complement activating. Finally, electrostatic charges on the membrane strongly promote platelet adhesion but confer favorable coagulation and complement-activating properties in the case of cationic charges.

Table 1. Density of surface-adherent mononuclear cells documented during contact time by light-microscopy: comparison between hydrogel (cuprophan) and microdomain structure (gambrane) polycarbonate-polyether membranes

Membrane	Coverage of the surface [1], %
Cuprophan[2]	30–40
Gambrane[3]	10–30

[1] Estimation from in situ LM observation after contact of 1×10^6 isolated mononuclear cells in serum-free RPMI 1640 for 30 min at 37 °C.
[2] Taken from Lundia IC, Gambro, Sweden.
[3] Taken from Lundia PRO, Gambro, Sweden.

One would predict that strong torque is exerted at the boundaries between hydrophobic and hydrophilic domains when cells (or plasma proteins) approach the surface. Our preliminary results [16] show that when cells sediment in a gravity field onto a cuprophan (hydrogel) membrane or a gambrane (polycarbonate-polyether) membrane with microdomain structure, significantly less mononuclear cells can be detected on the surface of the latter membrane. Gravitational forces must for obvious reasons be identical in the two experiments (table 1).

A difference in secondary membrane formation was also found as illustrated in figure 2, which compares fractions of proteins sequentially eluted from membrane surfaces and separated by SDS gradient gel electrophoresis (silver stain). Figure 2 shows the results for the hydrophobic (parent) and hydrophilic polyamide membranes indicating a remarkable reduction of hydrophobic protein-surface interactions by introduction of hydrophilic domains.

From the known characteristics of the polyamide membrane surface [17] one would predict that undesired hydrophobic interactions, causing platelet adhesion, have been avoided, as the size of the hydrophobic patches is below the critical threshold value. On the other hand, interaction of the hydrophilic PVP islands via hydrogen bonding should also be abrogated by the microdomain structure, since otherwise this should have led to complement activation (detailed discussion see below). In the following we wish to discuss whether these theoretical predictions are borne out by experimental findings in vitro and in vivo.

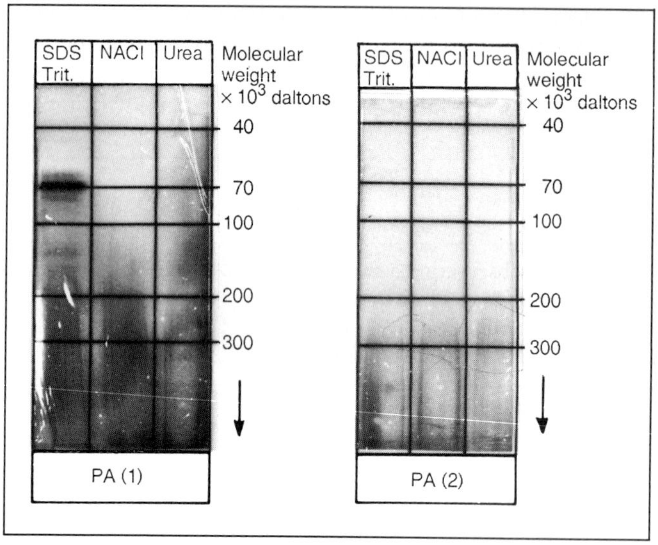

2

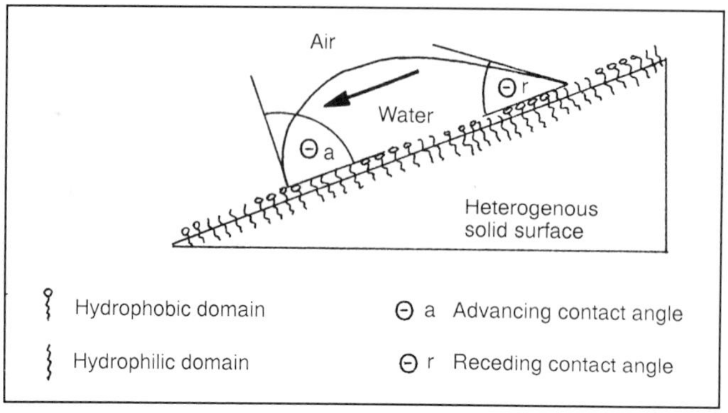

3

Fig. 2. Electrophoretic analysis (SDS-PAGE) of proteins sequentially eluted from polyamide membranes: comparison of the parent hydrophobic membrane (PA(1)) with the PVP-modified hydrophilic membrane (PA(2)). Membranes were exposed to heparinized plasma and extensively rinsed with Tris-buffered saline. The dialyzer was rinsed from the blood side and also by backfiltration from the dialysate compartment. Irreversibly bound protein fractions were sequentially eluted using Triton/SDS, 2 *M* NaCl and 2 *M* urea to break hydrophobic, ionic and hydrogen bonds, respectively. Protein patterns were visualised by highly sensitive silver stain techniques.

Fig. 3. Advancing and receding contact angle at the air/water/solid interface: schematic explanation for a hydrophilic-hydrophobic domain surface (orientation of hydrophobic molecules to the air interface: advancing is the air-water-solid interface; orientation of hydrophilic molecules to the water interface: receding is the air-water-solid interface).

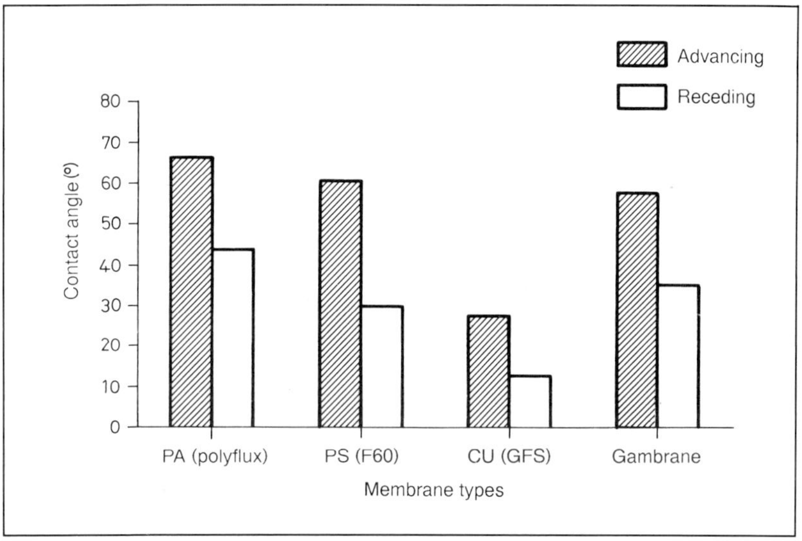

Fig. 4. Comparison of advancing and receding contact angles for different hollow fiber/flat sheet membranes. The measurements were performed in wet capillaries by monitoring pressure change during movement of the water-air-solid interface (PA = polyamide (polyflux); PS = polysulfone (F60); CU = cuprophan (GFS)) or using the 'tilt' plate method (30°) for flat sheets (gambrane). All data of contact angle ° are given as mean of n = 10 (SEM < 2°).

The Modified Polyamide Membrane – In vitro Surface Properties

Contact Angle

The contact angle at the interface air-water-solid surface is an index of wettability of a given surface and a measure of its hydrophilicity/hydrophobicity. If the surface is heterogenous, one will find a difference between the contact angle measured for advancing and receding contact. The difference (hysteresis) is a quantitative measure of dispersion of the surface properties [18]. This is explained in figure 3, which depicts the advancing and receding contact angle at a heterogenous surface (air-water-solid interface). For polyamide, a contact angle of 69° (advancing) and 38° (receding) was found by measuring at the air-water-solid interface of the hollow fiber 'skin' side. This compares with respective values of 28° and 13° for cuprophan, 58° and 35° for gambrane (polycarbonate-polyether) and 61° and 31° for polysulfone (fig. 4). Based on this observation and the fact of sur-

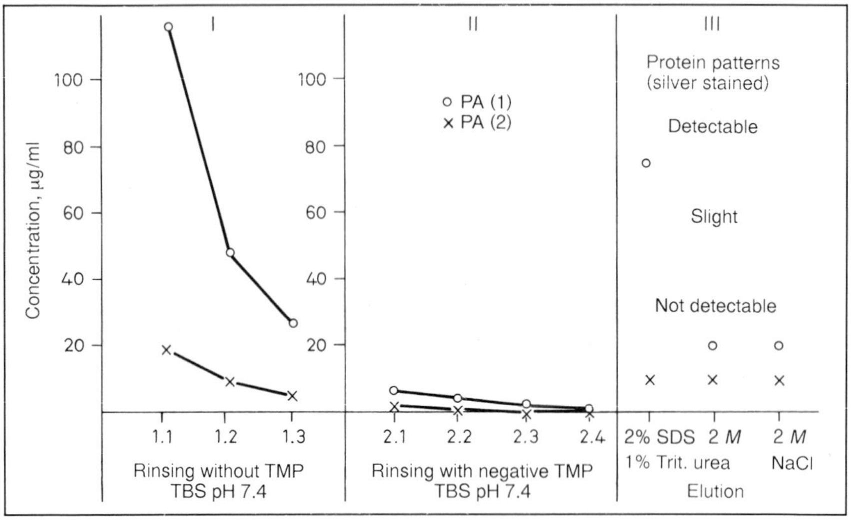

Fig. 5. Protein concentration in different rinsing cycles and elution steps of dialyzer membranes: comparison of the parent polyamide (PA(1)) and PVP-modified polyamide (PA(2) (polyflux)) membrane. Membranes were exposed to heparinized plasma. Protein concentrations were determined by the 'Lowry' technique in rinsing fractions (200 ml Tris-buffered saline (TBS) at 37 °C for 30 min and 1.3 m^2 membrane area at a flow rate of 200 ml/min and during backfiltration with a negative transmembrane pressure (TMP) of 100 mm Hg).

face smoothness of the 'skin', the conclusion can be drawn that marked heterogenicity exists with respect to the hydrophobic and hydrophilic character of the polyamide membrane surface.

Protein Adsorption

As outlined above, it can be predicted that protein adsorption to the polyamide membrane is minimal. This had been the rationale for the original development of this membrane. The prediction is well borne out by some qualitative and quantitative observations. Panichi et al. [19] examined the secondary protein layer on different membranes using an immunofluorescence technique. They compared cuprophan, polysulfone, polyacrylonitrile, polymethylmethacrylate and polyamide in contact with blood during dialysis sessions of 35 uremic patients. IgG, IgA and IgM were detected on all membranes with the exception of the polyamide membrane. This qualitative observation is in good agreement with our detailed

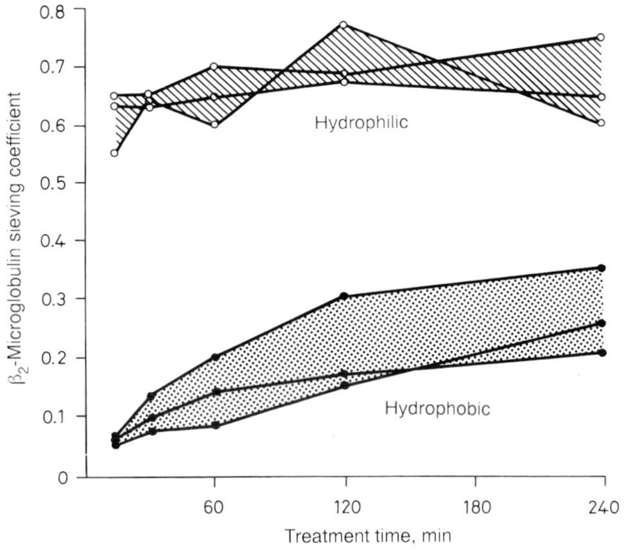

Fig. 6. Sieving coefficients for β_2-microglobulin as measured during hemofiltration with the hydrophobic parent polyamide membrane (PA1) and the modified hydrophilic polyamide (polyflux) membrane (PA2) in 3 patients. Taken from Flöge et al. [6].

quantitative studies. As demonstrated in figure 5, the total amount of protein which can be retrieved from the parent (hydrophobic) polyamide membrane and from the new hydrophilic polyamide membrane differs markedly and shows less protein adsorption to the hydrophilic version. This is true for the passive, loosely adsorbed layer (which can be retrieved by reversing the transmembrane pressure gradient) as well as for the protein layer adhering via hydrophobic interactions, hydrogen bonds or electrostatic interactions, respectively (fig. 2).

Minimal protein adsorption should lead to high sieving coefficients for proteins with an Einstein-Stokes radius below 3 nm, e.g. the 11.8-kD protein β_2m (Einstein-Stokes radius approximately 2.1 nm). In addition, one would predict that the sieving coefficient does not change with time. Both predictions are well borne out by the study of Flöge et al. [6] which is illustrated in figure 6. Hydrophilic as compared to hydrophobic polyamide membranes had quite superior performance characteristics.

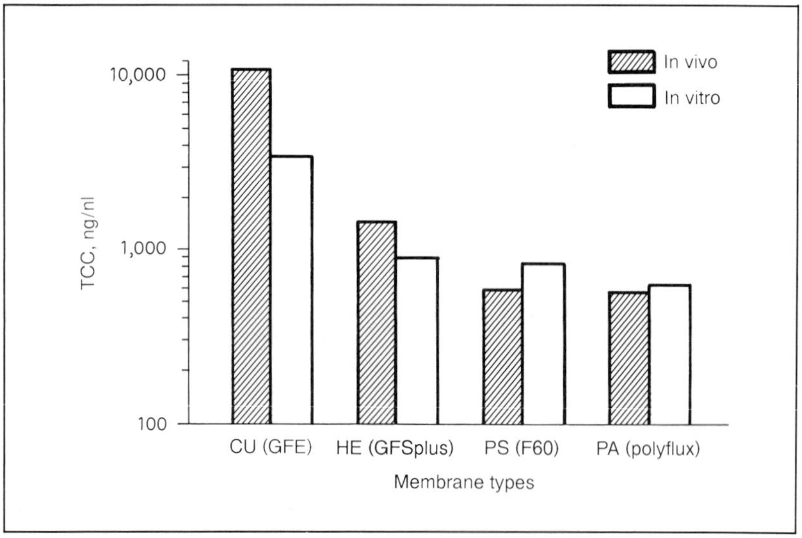

Fig. 7. Terminal complement activation during in vivo and in vitro contact of blood (or serum) with different membranes. In vivo data are given as the maximum arterial terminal complement complex (TCC) concentration during dialysis under standardized conditions. In vitro data represent the generation of TCC in a single path perfusion system using minidialyzers and freshly prepared human serum [21]: CU = cuprophan (GFE 120); HE = hemophan (GFSplus 120); PA = polyamide (polyflux 130); PS = polysulfone (F60); minidialyzers were prepared from commercially available dialyzers.

Complement Activation

Complement activation can be assessed by measuring either generation of the complement anaphylatoxins, i.e. C3a and C5a or their des-arg derivatives, respectively. In the past we have shown that measurement of the terminal complement complex (TCC) also provides a sensitive and highly specific index of complement activation by dialysis membranes [20, 21]. Using an ELISA we measured in vitro generation of TCC in minidialyzers produced from various membranes in identical geometrical configuration. As shown in figure 7, generation of TCC in vitro was highly predictive of the arterial TCC concentration when the same membranes were used for hemodialysis in vivo. In this context it is of note that TCC generation induced by the polyamide membrane is comparable to that by the high-flux polysulfone membrane and considerably less than that by cellulosic membranes.

The alternative pathway of the complement system is specifically geared towards distinguishing between 'self' and 'non-self' by the interaction of C3b, factor B and factor H with foreign surfaces. This readily explains why it constitutes the major pathway of dialysis membrane-induced complement activation. Recently, we found that the temporal stability of the C3-convertase (C3bBb) on different membrane surfaces is inversely related to the extent of surface-induced complement activation. This observation is compatible with the strictly limited interaction between the hydrophilic polyamide membrane surface and the alternative pathway proteins C3b and factor B [22].

Retention of Bacterial Products

When hydrophilic PVP domains were introduced into the hydrophobic polyamide membrane, it was a matter of great concern whether the desired retention and absorption properties of the hydrophobic membrane for bacterial cell wall products would be lost. This led to a number of detailed studies. Dinarello et al. [23] established that the parent polyamide membrane when used as an ultrafilter rejected IL-1- and TNF_{α}-inducing fragments of *Escherichia coli* and *Pseudomonas maltophilia.* Lonnemann et al. showed that similar fragments were effectively retained by polyamide dialyzers containing the modified membrane, when it was challenged in the dialysis mode using a Polyflux 130 dialyzer in a closed loop in vitro circuit [24]. Human mononuclear cells of healthy donors, when exposed to endotoxin-containing dialysate using cuprophan or cellulose acetate membranes released TNF_{α} into the supernatant, while they failed to do so when polyamide membranes were used under the conditions of the challenge. This observation illustrates that the desirable adsorptive properties via hydrophobic or other specific interactions are largely retained by the polyamide membrane with the heterogenous hydrophobic/hydrophilic microdomain structure.

Interaction with Blood Cells

Ogata et al. [25] examined various polyamides with different densities of hydrogen-bonds, hydrophilic sites and ionic groups. Adhesion of platelets was studied, using the microsphere column method. Platelet adsorption was intense when charged groups were present, considerably less adsorption was seen in the absence of charged groups.

Raij introduced the natural killer (NK) cell activity of peripheral blood lymphocytes (PBL) as an index of bioincompatibility, since NK cell

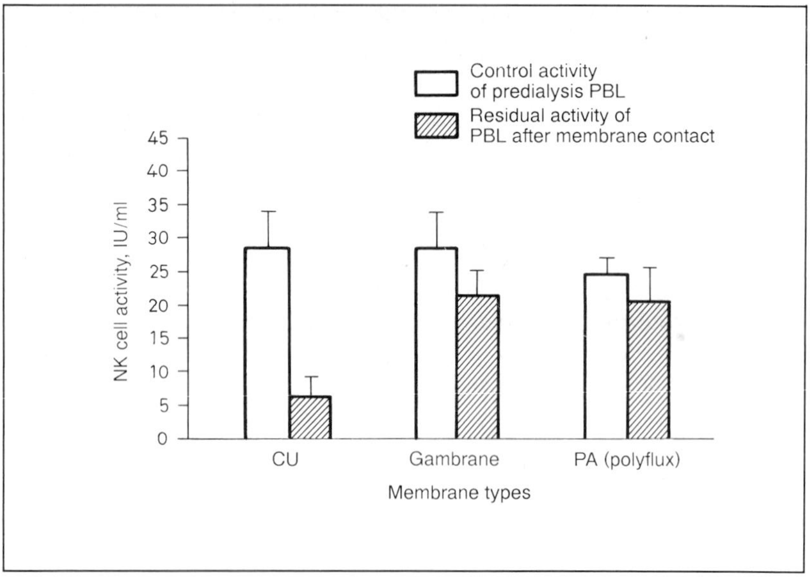

Fig. 8. Natural killer (NK) cell activity of human peripheral lymphocytes (PBL) after in vitro contact with dialyzer membranes. Data were taken from Kay and Raij [27] for cuprophan (CU), gambrane (polycarbonate-polyether) or kindly provided by Raij [26]. Data of NK cell activity of isolated PBL are given before and after 24 min contact of CU, gambrane ($\bar{x} \pm SD$, n = 3) and of PA ($\bar{x} \pm SD$, n = 4) using NK-sensitive cell line K562.

activity has been correlated with the resistance to viral infections and/or malignancies [26]. The effect of different membrane surfaces on NK cell activity in vitro is shown for isolated PBL in figure 8. Whereas PBL contact with cuprophan provokes a considerable suppression of NK cell function, only a minor effect has been observed for gambrane (polycarbonate-polyether) or polyamide membranes [26, 27].

In similar experiments we exposed isolated human mononuclear cells (MNC) under serum-free conditions in RPMI 1640 to the inner surface of capillary membrane modules (1×10^7 MNC in 2 ml medium for 4 h and a surface area of approximately 100 cm^2). After the incubation period, MNC were prepared for scanning electron microscopy. The view (fig. 9) into a polyamide hollow fiber further emphasizes that leukocytes do not interact with the microdomain structure of polyamide membranes.

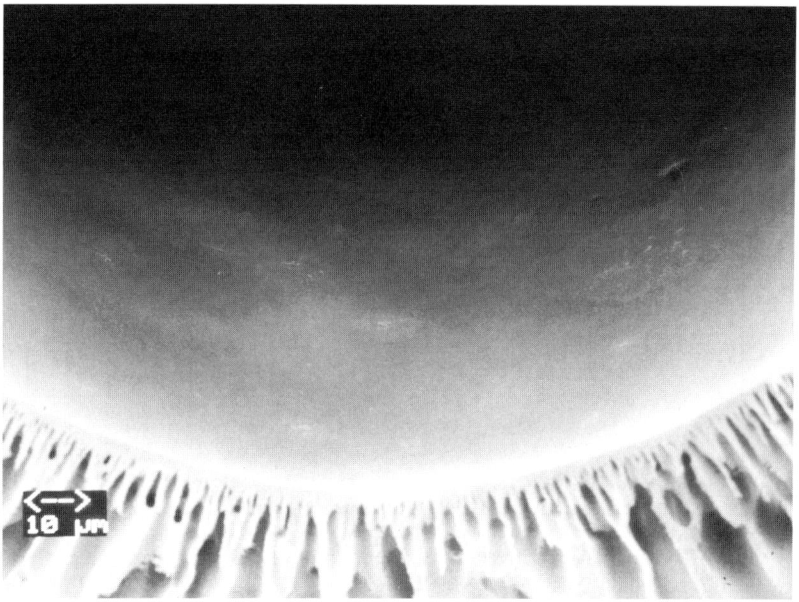

Fig. 9. Scanning electron micrograph of the inside of a polyamide (polyflux) fiber after contact with isolated mononuclear cells (MNC). MNC (1×10^7/ml) were incubated (4 h, 37 °C, RPMI 1640, serum free) on the inner side (blood side) of polyamide capilliaries in a mini module. After rinsing with saline, cells were fixed using glutardialdehyde and osmium tetroxide.

Biocompatibility of the Polyamide Membrane – In vivo Observations

Secondary Membrane Formation and Deposition of Cells

Von Bayer et al. [28] compared the surface morphology of capillary membranes in dialyzers and hemofilters. At the end of the dialysis treatment the membranes were examined after extensive rinsing and fixation using scanning electron microscopy. While intense fibrin layer formation with adhering blood cells was found with cuprophan and polymethylmethacrylate, no fibrin layer formation and only occasional thrombocyte adhesion were found with polyamide and polysulfone membranes. A point of note is the observation that polyacrylonitrile, a hydrogel structure, behaved similarly to cuprophan. This puzzling observation is easily explicable in the light of our introductory comments on the beneficial effects of the microdomain structure on cell adhesion.

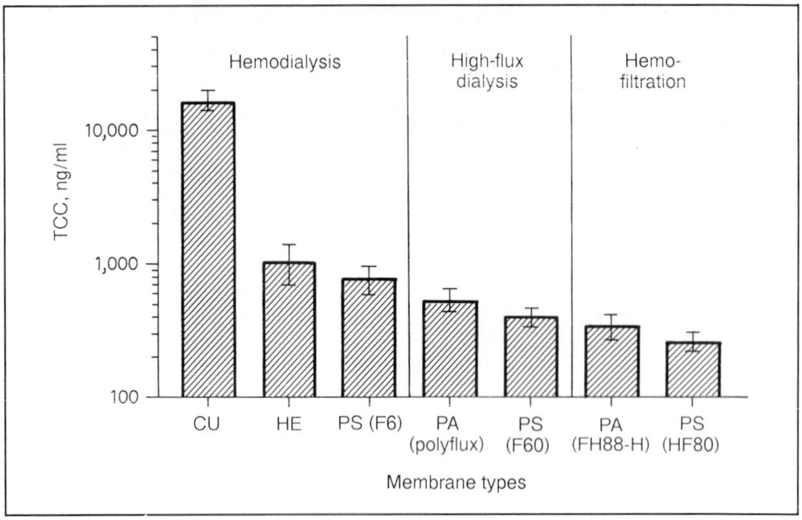

Fig. 10. Maximal arteriovenous difference of terminal complement complex (TCC) concentrations in plasma during hemodialysis, high-flux dialysis and hemofiltration [21]. Five patients randomly selected were treated in dialysis mode using CU (cuprophan), HE (hemophan) and PS (polysulfone; F6) (ETO-sterilized). In 2 further studies, 5 other patients were treated in high-flux dialysis and hemofiltration mode using PA (polyamide (polyflux 130, FH88H)) and PS (polysulfone (F60, HF80)) membranes. Data are calculated as $\bar{x} \pm SD$ of $((C_{out}-C_{in})_{max})$. Each group of patients was dialysed under identical conditions: HD = 1.2 m^2, 280 ml/min; HFD = 1.3 m^2, 250 ml/min; HF = 1.8 m^2, 500 ml/min.

Complement Activation

Activation of the complement system by dialysis membranes has been examined using the early anaphylatoxins (C3a and C5a or des-arg C3a and C5a, respectively). Measurement of the anaphylatoxins yields limited information since the blood concentration reflects the net result of the following processes: (i) surface-induced generation; (ii) removal by adsorption to the membrane; (iii) removal by filtration through the membrane, and (iv) removal by binding to receptors on circulating blood cells. Using a kinetic model Jorstad et al. [2] calculated net production of C3a during treatment with high-flux membranes. Even with the unmodified polyamide membrane (FH77) net production amounted only to 1,845 μg C3a and 6 μg C5a. Schäfer et al. [29] found low levels of C3a des-arg formation for both polyamide and polysulfone.

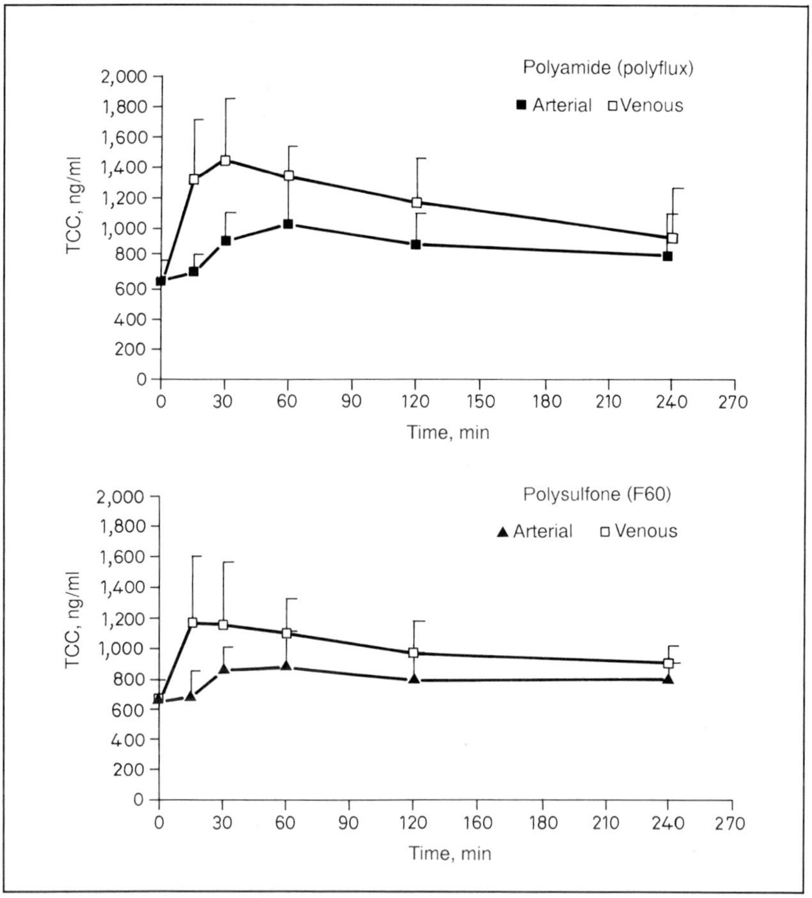

Fig. 11. Time courses of arterial and venous TCC concentration during high-flux dialysis with polysulfone (F60) and polyamide (polyflux 130) [31]. Data are given as $\bar{x} \pm$ SD for 5 patients treated under identical conditions of blood flow, heparinisation, dialysate flow and composition. Note that scale of the ordinate is expanded 10-fold relative to respective cuprophan data.

Using our ELISA for the terminal complement complex (TCC) we have examined generation of TCC with different dialyzer membranes [20]. Figure 10 shows the maximal arteriovenous difference of TCC (under standardized conditions of blood flow, heparinisation, etc.) studied as an intraindividual comparison with the same patients treated with the following membranes in capillary configuration: cuprophan, hemophan, polysulfone and polyamide. In the polyamide dialyzers very low TCC generation was noted and the generation tended to be even lower in hemofiltration mode, where blood flows up to 500 ml/min were used. The latter observation may be explained by the shorter contact time. Low TCC activation measured in vitro has been confirmed by Przchera et al. [30]. As shown in figure 11, the temporal profiles of TCC concentrations in arterial and venous lines are comparable when polyamide and polysulfone membranes were used. Net generation of TCC was consistently demonstrated.

At TCC concentrations of 1,500 ng/ml in the fluid phase, deposition of TCC onto cell membranes (and thus cell activation) will presumably take place, as recently shown by Salama et al. [31] and by ourselves [20]. We examined cell-surface TCC using fluorescent activated cell sorting (FACS) and immunostaining techniques (APAAP). Our studies showed preferential immunostaining for TCC on polymorphonuclear cells. This observation may explain the limited, but definite release of PMN-elastase or lactoferrin in patients treated with polysulfone or polyamide membranes (see below in fig. 15).

Thrombogenicity

Schultze et al. [32] scored capillary dialyzers for residual blood volume after extensive rinsing at the end of dialysis. Significantly better scores (1–2) were found for polyamide (Polyflux 130) as compared to hemophan (GFSplus 12) (2–3). Comparing heparin requirements, the authors noted identical dosage during treatment with both dialyzers. When the thrombin-antithrombin III complex (TAT) was measured by ELISA as a sensitive and reliable index of actual endogenous coagulation, less complex formation was found during treatment with polyamide capillary hemodialyzers as compared to hemophan capillary dialyzers (fig. 12).

Interaction with Thrombocytes and Leukocytes

As shown in figure 13, the polyamide membrane in capillary dialyzer configuration does not provoke a fall in platelet counts [33]. This observation has been confirmed by other investigators [1, 30, 32] and further illustrates minimal interaction with blood cells and explains the favorable thrombogenicity characteristics of the membrane.

In parallel, the drop in leukocyte counts which can be observed with cellulosic membranes is not found with polyamide membranes [1, 29, 30,

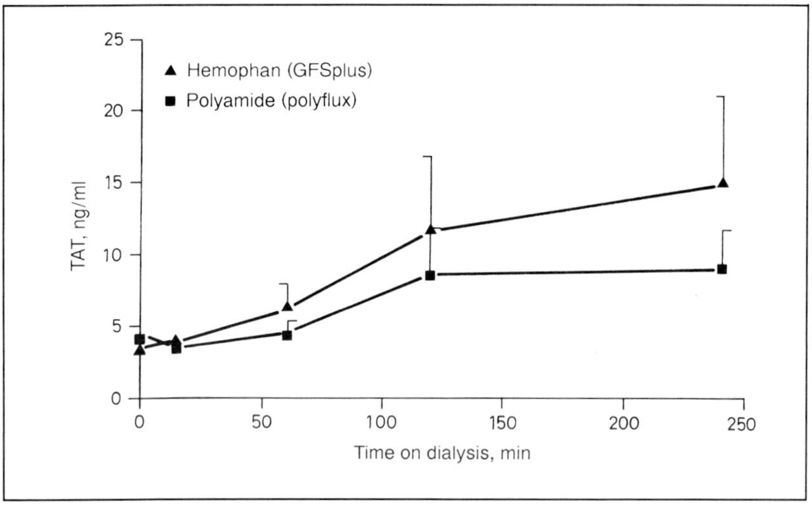

12

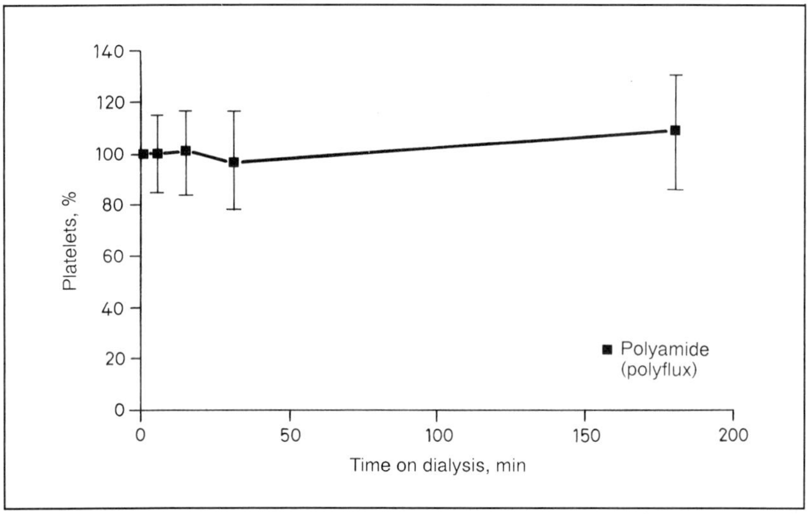

13

Fig. 12. Formation of the thrombin-antithrombin III-complex (TAT) during dialysis with polyamide and hemophan membranes. Taken from Schultze et al. [32]. Data are given as $\bar{x} \pm$ SE in a study comprising 10 patients, who were treated under identical conditions using polyamide (polyflux 130) and hemophan (GFSplus) membranes. TAT was measured by ELISA (Behring Co., FRG).

Fig. 13. Platelet counts in the course of one dialysis session from 14 different patients using dialyzers (polyflux 130) with polyamide membranes. Derived from Fournier and Morinière [33]. Data are given as $\bar{x} \pm$ SD expressed as percentage change from the starting value at time 0 min.

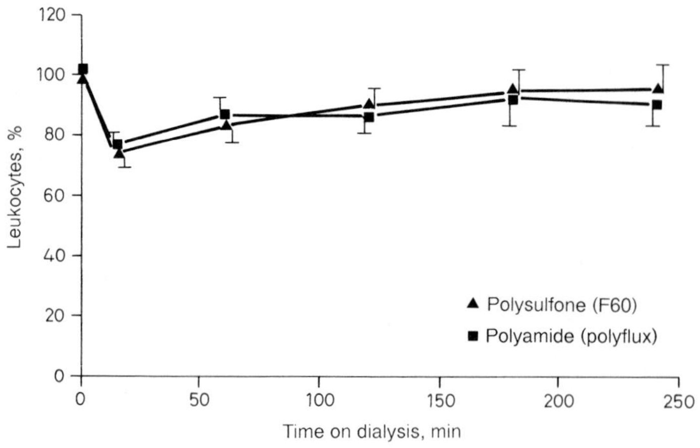

Fig. 14. Leukocyte counts ($\bar{x} \pm$ SE) in the course of dialysis session in 8 patients using dialyzers with polysulfone (F60) and polyamide (polyflux 130) membranes [29].

33, 34]. Figure 14 shows a minimal drop of peripheral leukocytes documenting a limited surface-induced cell activation via humoral mediators or via direct interaction with white blood cells [29].

Release of PMN Enzymes

Release of enzymes from PMN, e.g. elastase and lactoferrin, has been shown to be an important index of biocompatibility [35] indicating an altered granulocyte function as a consequence of the dialysis therapy. Figure 15 depicts the release of PMN-lactoferrin from the specific granules during dialysis treatment with different cellulosic and synthetic membranes [36]. The extent of lactoferrin release and the complement activation, measured as TCC generation (fig. 7), can be correlated for the membranes tested, except for polymethylmethacrylate. This membrane shows a remarkable increase of lactoferrin although only a minor complement activation has been reported. Release of PMN enzymes for different membranes was associated with alterations of further metabolic parameters of PMN, such as increasing glucose uptake and rise of intracellular Ca^{++} in resting and FMLP-stimulated PMN. Similar results, confirming the lowest

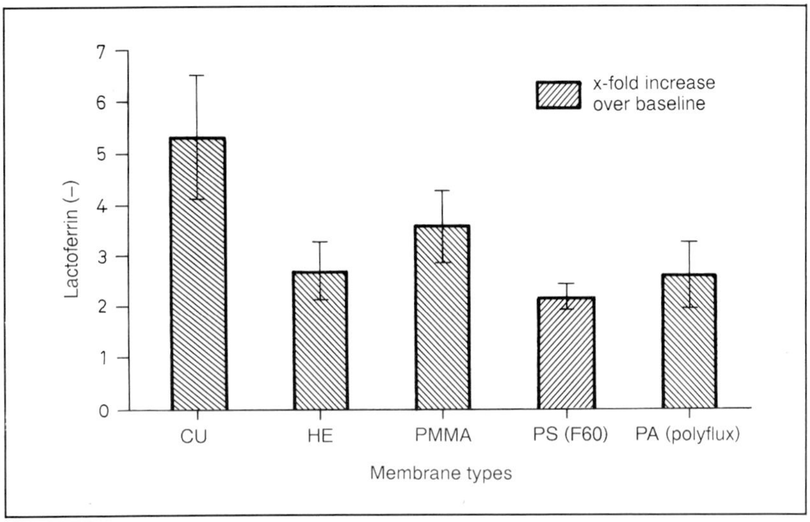

Fig. 15. Release of lactoferrin from polymorphonuclear cells in the course of dialysis. Data kindly provided by Hörl [36]. Data of n = 8 patients are given as the x-fold increase ($\bar{x} \pm$ SE) over predialysis baseline value. Patients were treated under identical conditions using ETO-sterilized cuprophan (CU), hemophan (HE), polymethylmethacrylate (PMMA), polysulfone (PS; F60) and polyamide (PA; polyflux 130) membranes.

PMN activation for polyamide and no significant difference in comparison to polysulfone, were obtained when release of elastase (from azurophilic granules) and release of lactoferrin were examined in an independent study [29].

Perspectives

The present analysis shows that the concept of hydrophobic-hydrophilic microdomain structure explains the favorable biocompatibility characteristics of the modified polyamide membrane. It has been possible to retain high adsorptive properties for endotoxins despite limited or no activation of blood cells and plasma proteins.

Optimization of the size of the hydrophilic domains will be a goal of the future. Today, the domains are relatively large (approximately 100–500 nm across) compared to the linear block-copolymer system used for the gambrane (polycarbonate-polyether) membrane. In the latter, the size of the hydrophilic domains is estimated to be smaller than 100 nm across. Whether a finer microtexture of the polyamide membrane will be possible by altering the conditions of membrane precipitation, or by introducing alternative polymers, will require further study.

In order to optimize nonthrombogenicity and to potentially achieve heparin-free dialysis, several solutions are conceivable, e.g. covalent modification of the membrane by oriented bonding of heparin [38]. This concept could possibly be further extended, e.g. by introducing active groups interacting with the enzymes of the coagulation cascade.

In the future the polyamide membrane may also find alternative uses. Apart from sterile filtration of fluids for medical purposes [39], further applications are conceivable, such as artificial lung, encapsulation of cells like hepatocytes or Langerhans cells, or use in biosensor systems. For sensing purposes polyamide membranes are particularly favorable, since transport distances for permeation of solutes are short in asymmetric membranes and polarisation due to secondary membrane formation is largely avoided with this membrane.

References

1 Von Herrath D, Hüfler M, Hartenstein K, Kutschera D, Schäfer K, Dati F: Good biocompatibility of the polyamide hemofilter. Blood Purif. 1988;6:106–110.

2 Jorstad S, Smeby LC, Balstad T, Wideroe TE: Generation and removal of anaphylatoxins during hemofiltration with five different membranes. Blood Purif 1988;6: 325–335.

3 Falkenhagen D, Brown GS, Thomanek W, Klinkmann H: Biocompatibility; in van Berlo A (ed): New Trends in Blood Purification. Eindhoven, University of Technology, 1986, pp 37–50.

4 Streicher E: Transport properties in filtration and dialysis membranes. Contr Nephrol. 1982;32:31–39.

5 Göhl H, Konstantin P, Gullberg CA: Hemofiltration membranes. Contrib Nephrol. Basel, Karger, 1982, vol 32, pp 20–32.

6 Flöge J, Granoleras C, Smeby L, Göhl H, Koch KM, Shaldon S: Hydrophilic high flux polyamide membranes for β_2-microglobulin removal. Trans Am Soc Artif Intern Organs 1987;33:309–311.

7 Buck R, Göhl H: A permselective asymmetric membrane suitable for hemodialysis and for manufacturing said membrane. Eur Patent Application, No 0305 787 A1, 1988.

8 Okano T, Kataoka K, Sakurai Y, Shimada M, Miyahara M, Akaike T, Shinohara I: Molecular design of block and graft copolymers having the ability to suppress platelet adhesion. Artif Organs 1981;5(suppl):468–470.

9 Yui N, Tanaka J, Sanui K, Ogata N, Kataoka K, Okano T, Sakurai Y: Characterization of the microstructure of poly(propylene oxide)-segmented polyamide and its suppression of platelet adhesion. Polymer J 1984;16:119–128.

10 Vesely D, Finch DS: Microstructure of miscible polymer blends. Macromol Chem 1988;16:329–340.

11 Kumpf RJ, Wehrmann R, Karbach A, Köhler B, Nerger D, Pielartzik H: A novel route to poly(arylether(-thioether))/LC-polycondensate block and graft copolymers. Book of abstracts, Makromolekulares Kolloquium, Freiburg, 1991, p 19.

12 Betz M, Hänsch GM, Rauterberg EW, Bommer J, Ritz E: Cuprammonium membranes stimulate interleukin-1 release and arachidonic acid metabolism in monocytes in the absence of complement. Kidney Int 1988;34:67–73.

13 Ziats NP, Miller KM, Anderson JM: In vitro and in vivo interactions of cells with biomaterials. Biomaterials 1988;9:5–13.

14 Brash JL, Horbett TA (eds): Proteins at inter-faces: Physicochemical and biochemical studies. American Chemical Society Symp Ser 343, Washington, 1987, p 343.

15 Matsuda T: Biological responses at non-physiological interfaces and molecular design of biocompatible surfaces. Nephrol Dial Transplant 1989;4(suppl):60–66.

16 Stier E, Deppisch R, Nawroth PP, Bierhaus A, Hänsch GM, Ritz E: Release of mediators and activation of protooncogenes in mononuclear cells is dependent on the physicochemical structures of dialysis membranes (abstract). J Am Soc Nephrol 1990;1:379.

17 Strathmann H, Göhl H: Membranes for blood purification: State of the art and new developments. Contrib Nephrol. Basel, Karger, 1990, vol 78, pp 119–141.

18 Andrade JD, Smith LM, Gregonis DM: The contact angle and interface energetics; in Andrade JD (ed): Surface and Interfacial Aspects of Biomedical Polymers. New York, Plenum Press, 1985, vol 1 (Surface Chemistry and Physics), pp 249–292.

19 Panichi V, Bianchi AM, Casarosa L, Cirami C, Parrini M, Grazi G, Montagnani E, Neri M, Palla R: Anaphylatoxin generation and protein layer with five different dialysis membranes (abstract). Blood Purif 1989;7:266.

20 Deppisch R, Schmitt V, Bommer J, Hänsch GM, Ritz E, Rauterberg EW: Fluid phase generation of terminal complement complex as a novel index of bioincompatibility. Kidney Int 1990;37:696–706.

21 Deppisch R, Schmitt V, Bommer J, Risler T, Schaefer K, Ritz E, Rauterberg EW: Complement activation by conventional and highly permeable membranes – studies with micro-ELISA for terminal complement complexes (TCC). Kidney Dial 1989; 27:64–67.

22 Deppisch R, Rittinger K, Han KT, Reichel M, Vienken J, Ritz E, Rauterberg EW: Factors regulating the degree of 'foreign' surface-induced activation of the complement system: role of factor H, factor B and C3b. Book of abstracts, 4th Eur Symp on Biomaterials, Siena, 1991, p 64.

23 Dinarello CA, Lonnemann G, Maxwell R, Shaldon S: Ultrafiltration to reject human interleukin-1-inducing substances derived from bacterial cultures. J Clin Microbiol. 1987;25:1233–1238.

24 Lonnemann G, Flöge J, Schindler R, Behme T, Lenzner B, Shaldon S, Koch KM: Passage of cytokine-inducing E. coli fragments (CEF) through various hemodialysis (HD) membranes (abstract). Kidney Int 1989;35:254.

25 Ogata N, Sanui K, Tanaka H, Takahashi Y, Sakurai Y, Akaike T, Okano T, Kataoka K: Cellular affinity of polyamides having various functional groups. I. Platelet adhesion. J Appl Polym Sci 1981;26:2293–2303.
26 Raij L: Personal commun.
27 Kay NE, Raij L: Differential effect of hemodialysis membranes on human lymphocyte natural killer function. Artif Organs 1987;11:165–167.
28 Von Bayer H, Lajous-Petter A, Debrand A, Herbst R: Untersuchungen der Blutkontaktfläche von Dialyse- und Hemofiltrationsmodulen mit Raster- und Transmissionselektronenmikroskopie: Die mögliche Rolle der Membrane bei der Entstehung der Dialyse-assoziierten Amyloidose. Nieren Hochdruckkr 1987;16:185–194.
29 Schäfer RM, Gilge U, Göhl H, Heidland A: Evaluation of a new polyamide membrane (polyflux 130) in high-flux dialysis. Blood Purif 1990;8:23–31.
30 Przchera M, Rothfuss C, Deppisch R, Rauterberg EW, Risler T: Der terminale Komplementkomplex SC5b-9 als Index für die Biokompatibilität von Highflux-Dialysatoren. Abstract, 19. Kongress der deutschen Gesellschaft für Nephrologie. Nieren Hochdruckkr 1988;9:351.
31 Salama A, Hugo F, Heinrich D, Höge R, Müller R, Kiefel V, Müller-Eckhardt C, Bhakdi S: Deposition of terminal C5b-9 complement complexes on erythrocytes and leukocytes during cardiopulmonary bypass. N Engl J Med 1988;318:408–414.
32 Schultze G, Göhl H, Hollmann S, Sinah P: Formation of thrombin-antithrombin complex using polyamide and cellulosic dialyzers. Book of abstracts, Eur Soc Artificial Organs, 1991.
33 Fournier A, Morinière P: A new capillary hemodiafilter based on polyamide membrane: polyflux 130. Clinical report. Lund, Gambro, 1989, HCE 7497.
34 Schäfer K, von Herrath D, Hüfler M: Stable pO_2 during hemofiltration in spite of a drop in leucocytes (letter). Nephron 1982;32:377.
35 Hörl WH, Steinhauer HB, Riegel W, Schollmeyer P, Schäfer RM, Heidland A: Effect of different dialyzer membranes on plasma levels of granulocyte elastase. Kidney Int 1988;33(S24):S90–S91.
36 Hörl WH: Personal commun.
37 Bommer J, Ritz E: Ethylene oxide as a major cause of anaphylactic reactions in dialysis. Artif Organs 1987;11:111–117.
38 Engbers GH; Feijen J: Current techniques to improve the blood compatibility of biomaterial surfaces. Int J Artif Organs 1991;14:199–215.
39 Erley CM, von Herrath D, Hartenstein-Koch K, Kutschera D, Amir-Moazami B, Schäfer K: Easy production of sterile, pyrogen-free dialysate. Trans Am Soc Artif Intern Organs 1988;34:205–207.

Professor Dr. E. Ritz, Abteilung Innere Medizin, Krehl-Klinik,
Bergheimerstrasse 58, D-W–6900 Heidelberg (FRG)

Shaldon S, Koch KM (eds): Polyamide – The Evolution of a Synthetic Membrane for Renal Therapy. Contrib Nephrol. Basel, Karger, 1992, vol 96, pp 47–63

Pyrogen Retention by the Polyamide Membranes

Gerhard Lonnemann[a], *Arezki Mahiout*[a], *Ralf Schindler*[a], *Clark K. Colton*[b]

[a]Department of Nephrology, Medizinische Hochschule Hannover, FRG;
[b]Massachusetts Institute of Technology, Cambridge, Mass., USA

Pyrogens and Pyrogen Assays

Endotoxins and exotoxins are substances released from gram-negative and gram-positive bacteria with a wide distribution of molecular weight (0.2–1,000 kD). Because of their ability to induce acute-phase reactions including fever these substances are often referred to as pyrogens. This group of bacterial products is very heterogeneous but can be divided into two groups. Group 1 comprises cell wall components of gram-negative bacteria such as endotoxins otherwise known as lipopolysaccharides (LPS), lipid A, peptidoglycans and muramyl peptides. These products are released from bacterial debris. Group 2 consists of bacterial exotoxins; these substances are proteins and derived polypeptides which are actively secreted from living bacteria.

The intact LPS molecule (> 100 kD) and the peptidoglycan polymer (> 60 kD) are large molecules which are likely to be rejected by any dialysis membrane. However, muramyl peptides (0.4–1 kD), the natural occurring subunits of peptidoglycans, as well as lipid A (2–4 kD in its monomeric form), the very pyrogenic subunit of LPS, are substances of low molecular weight. These products may not even be rejected by low-flux membranes such as cuprophan. Micelles which are formed from the hydrophobic lipid A in aqueous solutions may be small enough to cross high-flux membranes such as polyamide or polysulfone. Besides these cell wall components of gram-negative bacteria, it has recently been reported that gram-negative water bacteria, *Pseudomonas* species in particular, also produce exotoxins [1]. Furthermore, there is evidence that *Pseudomonas* species release pyrogenic peptide-like fragments of molecular weights as low as 0.2–5 kD [2].

It is well known that microorganisms grown in rich culture medium produce pyrogenic substances. However, pyrogens are also released from bacteria grown in nutrient-poor media such as water, saline, infusion fluids, or dialysate [3]. In order to test whether material intended for parenteral use is pyrogen-free, the rabbit pyrogen assay and the *Limulus* amoebocyte lysate (LAL) test are used routinely. However, there are limitations to both methods. The rabbit pyrogen assay is not sensitive enough to prove the absence of pyrogens; and the LAL test, although very sensitive for the detection of LPS and lipid A (detection limit approximately 1 pg/ml), fails to detect other pyrogens such as exotoxins and muramyl peptides.

The acute-phase changes induced by pyrogens are mediated by cytokines such as interleukin-1 (IL-1) and tumor necrosis factor (TNFα) [4], which are predominantly produced by circulating mononuclear cells (MNC). Mononuclear cells separated from whole blood using centrifugation can be transferred into cell cultures and incubated with contaminated fluids to determine the amount of cytokine induction produced by MNC from contaminated fluids. Compared to the LAL and the rabbit pyrogen test the in vitro incubation of human MNC to measure cytokine production is a more relevant and biologically meaningful test. Several studies have shown that MNC produce significant amounts of IL-1 or TNFα in response to small amounts of both LAL-positive and LAL-negative bacterial substances, including LPS and lipid A [5] and muramyl peptides [6]. A study comparing the IL-1-inducing activity of various types of synthetic lipid A molecules, peptidoglycans and muramyl peptides demonstrated that lipid A is the most active cell wall component of gram-negative bacteria [7].

It has been consistently shown that LPS concentrations as low as 50 pg/ml induce significant amounts of IL-1 in MNC. The advantage of the MNC test system to detect pyrogens which are not detected by the LAL test seems to be of increasing importance, as gram-negative water bacteria, e.g. *Pseudomonas* species have recently been shown to release very active TNFα-inducing substances of low molecular weight which are negative in the LAL test [2]. The relevance of the MNC test system in measuring pyrogens is reinforced by the correlation of in vitro cytokine induction from MNC for recombinant growth hormone preparations and the incidence of febrile reactions in vivo seen with contaminated lots which were negative to LAL testing [8].

We used the in vitro incubation of human donor MNC in combination with various physical processes to challenge the polyamide membranes for their capacity to reject pyrogenic substances.

Chemical Composition of the Polyamide Membranes and Mechanisms Involved in the Retention of Pyrogens

The parent polyamide is an asymmetric membrane with a thickness of approximately 60 µm and a molecular weight cut-off of approximately 40 kD. This synthetic membrane is intensely hydrophobic and is today only used in ultrafilters. It has a skinned finger structure and a high water flux. The modified membrane used for hemodialysis and hemodiafiltration (Polyflux dialyzer series) and hemofiltration (FH77H and FH88H) is a copolymer consisting of hydrophilic polyvinyl pyrrolidone (PVP) domains in the hydrophobic polyamide membrane. PVP is admixed with the polyamide during production of the membrane in order to render it wettable on contact with water. However, the surface of the membrane is not completely covered by PVP leading to the presence of hydrophobic patches. The hydrophobic domains are thought to be important in the interaction of the polyamide membrane with pyrogenic bacterial substances.

Because of the wide variation in molecular weight of bacterial substances of low molecular weight, rejection by size exclusion is unlikely to be an effective mechanism to retain pyrogens with the polyamide membranes. Other mechanisms involved in the interaction of the membrane polymer with bacterial substances include binding or adsorption. Increased hydrophobicity of a membrane has been shown to increase the affinity for adsorption of proteins [9, 10]. It may also favor adsorption of the hydrophobic, long-chain acyl moieties of lipid A [7] as well as of other hydrophobic moieties such as aromatic side chains on certain amino acids which may be present in peptide-like fragments derived from exotoxins [1].

The precise mechanism(s) responsible for adsorption of pyrogens to polyamide are unknown and undoubtedly complex. However, it is instructive to examine the basic chemical structure of the polyamide polymer (fig. 1). Polyamide contains benzene rings which may contribute to its

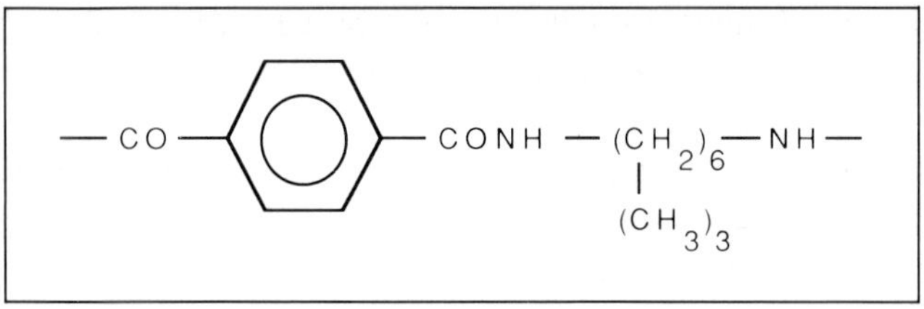

Fig. 1. The basic chemical compositional element of the polyamide polymer.

adsorptive capacity in two ways. First, they contain conjugated double bonds involving π-orbitals in which electrons are not localized to a particular site but are free to move over the entire length of the bond system. These high-mobility, easily displaceable electrons are thought to be responsible for the unusually large energies of interaction between large molecules containing conjugated bonds [11]. Secondly, in common with other hydrophobic moieties, they participate in the so-called hydrophobic interaction, a primarily entropic phenomenon that accounts for the unusually strong interaction between nonpolar molecules and surfaces in water [12]. This interaction arises from the hydrophobic effect, which is associated with restructuring of water around nonpolar surfaces to which the water molecules cannot form hydrogen bonds. When hydrophobic surfaces come together, removal of this ordered water leads to a large favorable increase in entropy. This entropy gain drives protein adsorption on solid surfaces, when hydrophobic parts of the protein and of the adsorbent surface interact [13], and it may be the predominant mechanism influencing adsorption of pyrogens to the polyamide membrane.

The Polyamide Ultrafilter Membrane in Water Treatment: Ultrafiltration to Reject Cytokine-Inducing Substances Derived from Bacterial Cultures

In order to test the ability to reject pyrogens derived from bacterial cultures, the polyamide ultrafilter (U 2000) was tested in ultrafiltration experiments [14]. Pyrogens were measured by their activity to induce IL-1 production in human MNC cultured for 24 h. IL-1 was measured in MNC supernatants using a bioassay of the IL-1-induced proliferation of murine thymocytes [15] or of a murine cell line [16] in costimulation with the mitogen phytohemagglutinin. Murine T cell proliferation was determined after 48 h of incubation measuring the uptake of [^{3}H]thymidine in cell lysates by liquid scintillation. In this study, human MNC consistently produced significant amounts of IL-1 in response to 50 pg/ml of the reference RE-5 endotoxin derived from *Escherichia coli.*

Bacterial Culture Filtrates

Two bacterial culture filtrates were used to challenge the polyamide membrane. A clinical isolate of *E. coli* was grown in brain heart infusion broth, a water contaminant of the genus *Pseudomonas* was grown in trypticase soy broth. Both bacterial cultures were performed in a shaking water bath at 37 °C for 48 h. The number of organisms after incubation was approximately 5×10^8/ml for all cultures. The broths were centrifuged and filtered (pore size 0.22 μm) to produce bacterial filtrates which were proven to be sterile in subsequent cultures for 48 h. The endotoxin concentration as measured by the LAL assay was approximately 100 μg/ml in the culture filtrates of *E. coli* and *Pseudomonas.*

Ultrafiltration with the Polyamide Ultrafilter U 2000

The polyamide ultrafilter (U 2000) was connected to pyrogen-free blood tubing sets under aseptic conditions and rinsed with 2 liters of pyrogen-free saline. The outlet port of the ultrafilter was then clamped and 2 more liters of saline were ultrafiltered. Two sterile, pyrogen-free 4-liter flasks were employed as the supply and drainage vessels (fig. 2). Sterile pipettes were connected to the tubings to drain the supply flask. The peristaltic pump interposed between the supply vessel and the inlet port of the filter maintained a flow rate of 200 ml/min for all ultrafiltration experiments. The ultrafilter was challenged with different dilutions of the bacterial culture filtrates in 4 liters of saline. After the rinsing procedure and at the end of each 4-liter ultrafiltration, samples were taken directly from the inlet and the ultrafiltrate outlet port. Samples were incubated with human MNC to measure IL-1-inducing activity.

Rejection of IL-1-Inducing Substances Derived from Pseudomonas *Culture Filtrate*

The ultrafilter was first challenged with 4 liters of saline and then with successive 4-liter portions of *Pseudomonas* culture filtrate. When challenged with the *Pseudomonas* filtrate, which induced IL-1 from MNC at a

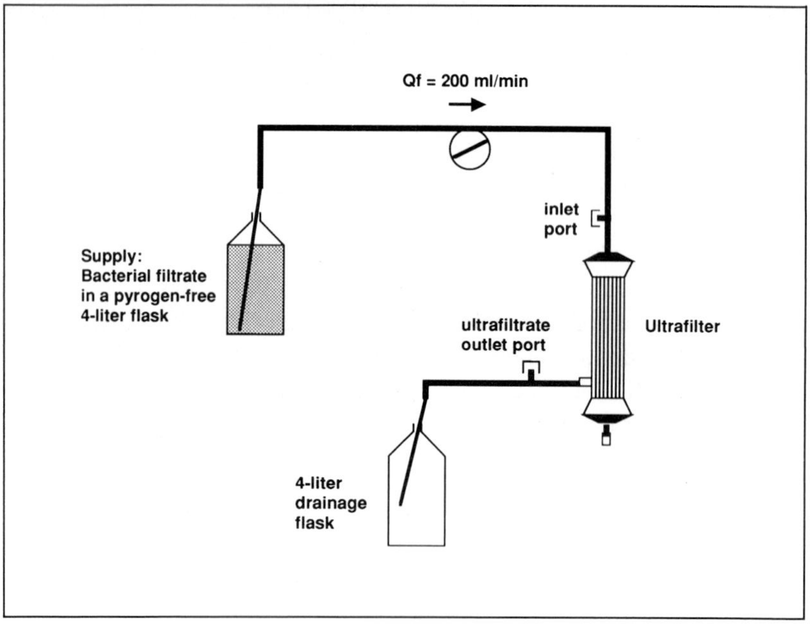

Fig. 2. The system used for the ultrafiltration experiments with polyamide ultrafilters (U 2000). Derived from Dinarello et al. [14].

dilution of 10^{-3} in the prefilter sample, there was no detectable IL-1-inducing activity in postfilter samples from 4 through 32 liters of ultrafiltrate (fig. 3). In further experiments the ultrafilter was challenged with total volumes of 32 liters containing various concentrations of *Pseudomonas* filtrate in saline (fig. 4). Prefilter dilutions of 10^{-4} induced large amounts of IL-1. However, postfilter samples taken after 32 liters only induced significant IL-1 production when the challenge dilution was 10^{-2}. Thus, the rejection factor of the ultrafilter for *Pseudomonas* culture filtrate was 100.

Rejection of IL-1-Inducing Substances Derived from E. coli *Culture Filtrate*

Using the same experimental design the polyamide ultrafilter was challenged with various dilutions of a culture filtrate derived from *E. coli.* The *E. coli* filtrate was very potent in inducing IL-1 from MNC. Pre- and

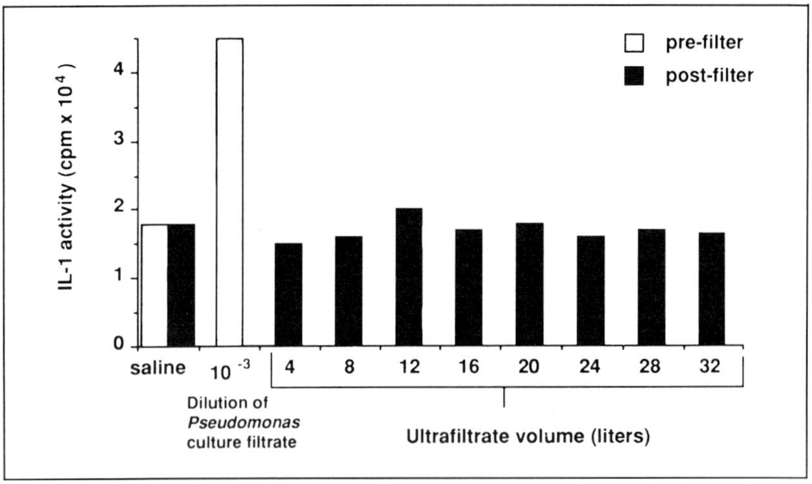

3

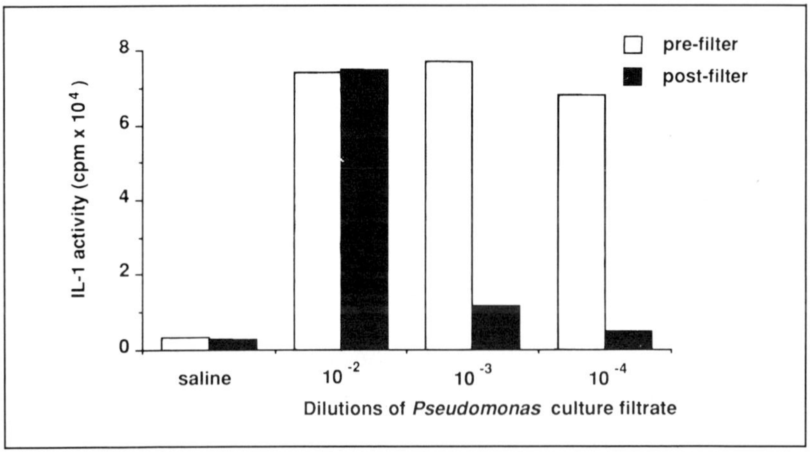

4

Fig. 3. IL-1-inducing activity in pre- and post-filter samples taken from the polyamide ultrafilter (U 2000) challenged with a *Pseudomonas* culture filtrate. Four-liter portions adding up to a total volume of 32 liters of a 10^{-3} dilution of the *Pseudomonas* filtrate in saline were pumped through the filter. 'Saline' indicates control samples taken after the rinse of the system. IL-1 activity (vertical axis) is expressed as counts per minute (cpm) of [^{3}H]thymidine uptake by murine thymocytes in the IL-1 bioassay. Derived from Dinarello et al. [14].

Fig. 4. IL-1-inducing activity in pre- and post-filter samples taken after 32 liters of ultrafiltrate. The U 2000 filter was challenged with a *Pseudomonas* culture filtrate in the dilutions indicated under the horizontal axis. 'Saline' indicates control samples taken after rinsing the system with 4 liters of saline. IL-1 activity is expressed as cpm. Derived from Dinarello et al. [14].

postfilter samples were taken after 8-liter portions of ultrafiltrate. As depicted in figure 5, prefilter dilutions of 10^{-2} to 10^{-9} induced large amounts of IL-1. The ultrafiltrate only contained IL-1-inducing substances when the dilution of the challenge solution was 10^{-4} or less. Consequently, the rejection factor of the polyamide ultrafilter for *E. coli*-derived IL-1-inducing substances was 100,000.

These data indicate that ultrafiltration through the polyamide ultrafilter removes pyrogenic substances derived from gram-negative bacteria very effectively. There is a wide margin of safety with the observed rejection factors of 100–100,000, depending on the specific bacterial filtrate used to challenge the filter. The dilution of the bacterial filtrates resulting in IL-1-inducing activity in the postfilter samples was 10^{-2}, which represents approximately 5×10^6 organisms per ml or 1 µg/ml LAL active material. These are rather high numbers compared to what one would expect to find in aqueous solutions such as dialysate which are contaminated with water bacteria. With lower degrees of contamination in the challenge fluids, ultrafiltration through the polyamide ultrafilter rejected IL-1-inducing bacterial substances from a total volume of 32 liters of saline. According to these results the estimated maximal amount of pyrogens rejected by the polyamide ultrafilter is between 3.2 and 32 mg of endotoxin as measured by the LAL assay.

Adsorption of Pyrogens to the Polyamide Membrane

In order to further investigate the question of whether rejection of bacterial substances by the polyamide membrane is due to molecular size exclusion, adsorption or both, additional experiments were performed. First, ultrafiltration was done in the in vitro system described in figure 2 with polyamide ultrafilters which were damaged by cutting a hole in a single hollow fiber. These damaged filters failed the so-called bubble test for intact filters. As shown in figure 6, challenge of the filter with a 10^{-2} dilution of the *E. coli* culture filtrate resulted in the passage of IL-1-inducing material at the end of 8 liters of ultrafiltrate, but remarkable rejection was apparent at lower challenge concentrations. These data suggest that adsorption of IL-1-inducing bacterial substances to the polyamide membrane occurs.

Adsorption of cytokine-inducing substances derived from cultured bacteria to the polyamide membrane was further studied in experiments

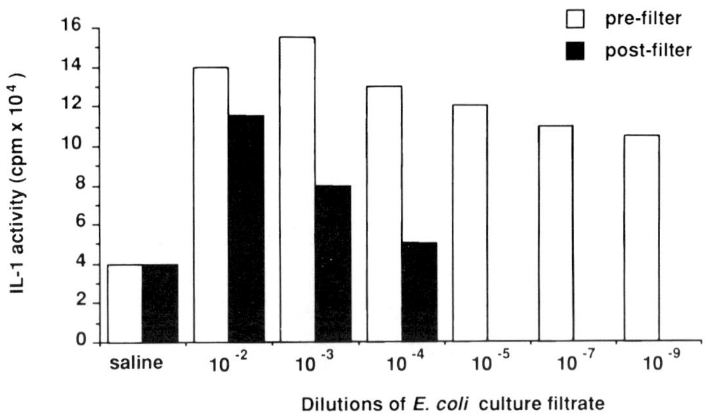

5

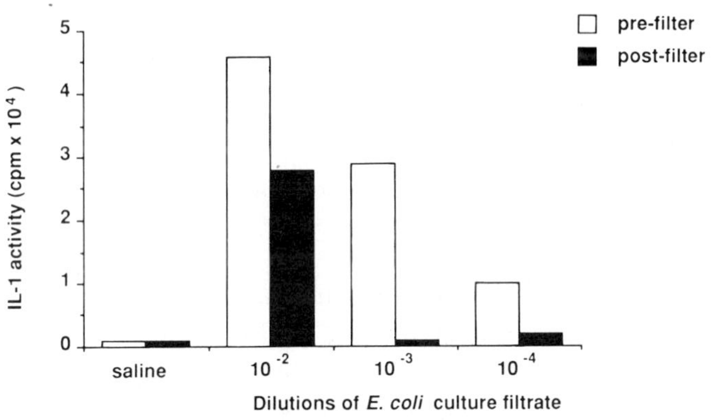

6

Fig. 5. IL-1-inducing activities in pre- and post-filter samples taken from a polyamide ultrafilter challenged with various dilutions of a bacterial filtrate derived from cultured *E. coli.* Samples were taken after 8-liter portions of ultrafiltrate. The dilutions of the *E. coli* filtrate used are indicated under the horizontal axis. IL-1 activity is expressed as cpm of incorporated [^{3}H]thymidine into murine thymocytes in the IL-1 bioassay. Derived from Dinarello et al. [14].

Fig. 6. IL-1 activity produced by human MNC incubated with pre- and post-filter samples of a damaged polyamide ultrafilter challenged with *E. coli* culture filtrate dilutions. For further details see figure 5.

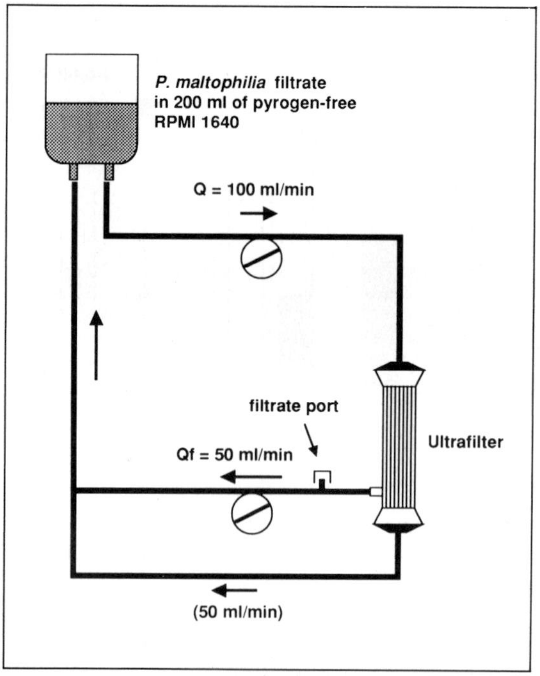

Fig. 7. Flow diagram of closed loop ultrafiltration. The flow rates and flow directions are indicated in the figure. Samples were taken from the bag before and after 30 min of recirculation as well as from the filtrate port after 30 min.

using a different design, described in figure 7. The in vitro circuit was aseptically assembled using either a U 2000 ultrafilter containing the parent polyamide membrane or a special ultrafilter containing the PVP-modified polyamide membrane and blood tubing sets. After rinsing with two liters of pyrogen-free saline the system was air-drained under sterile conditions before a bag containing a culture filtrate of *Pseudomonas maltophilia* diluted in 200 ml of pyrogen-free tissue culture medium (RPMI 1640) was connected. A pump interposed between the bag and the filter-inlet maintained a flow rate of 100 ml/min. A second pump connected to the filtrate line was set at an ultrafiltration rate of 50 ml/min. The filtrate line was reconnected to the filter outlet line to complete the closed loop circuit. Three dilutions of the *P. maltophilia* filtrate in RPMI (4×10^{-1}, 8×10^{-1} and 10^{-2}) were used to challenge the membrane. Samples were taken from the bag before and after 30 min of recirculation as well as from

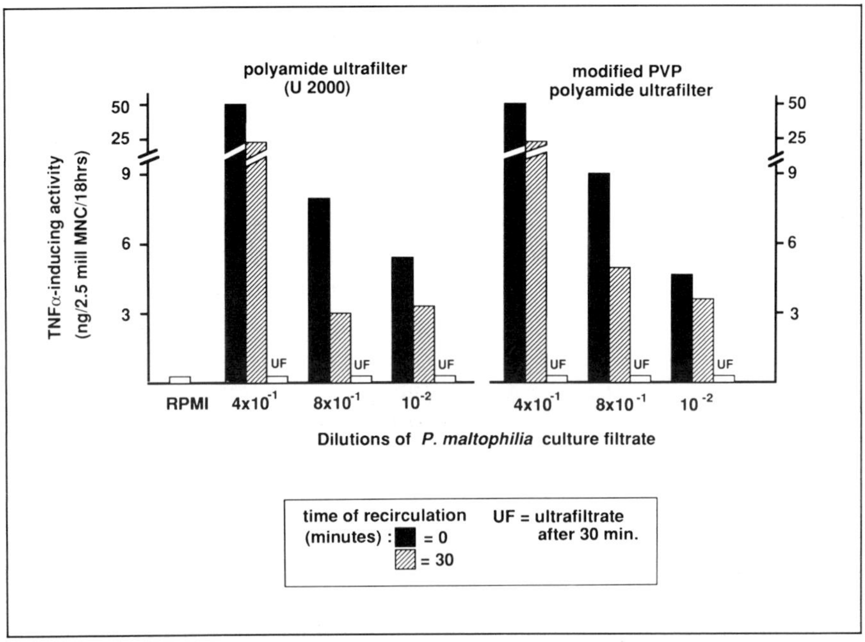

Fig. 8. TNFα-inducing activity derived from cultured *P. maltophilia* in the closed loop ultrafiltration system before and after 30 min of recirculation. Results for the polyamide ultrafilter (U 2000) are shown on the left panel, those for the modifed PVP polyamide ultrafilter are shown on the right. Dilutions of the *P. maltophilia* filtrate are indicated under the horizontal axis. 'RPMI' indicates control MNC incubations with tissue culture medium alone. TNFα-inducing activity is expressed as ng per 2.5×10^6 MNC per 18 h of incubation.

the filtrate port after 30 min. All samples were incubated with human MNC for 18 h. Following incubation, TNFα was measured in MNC supernatants by radioimmunoassay [17]. TNFα-inducing activity was expressed in ng/2.5×10^6 MNC/18 h. The results depicted in figure 8 show that both the regular ultrafilter membrane (U 2000) as well as the special more hydrophilic membrane of the same surface area adsorb 40–60% of the *P. maltophilia*-derived substances under the applied in vitro conditions. However, with the lower challenge concentrations of 8×10^{-1} and 10^{-2}, the parent hydrophobic ultrafilter membrane U 2000 showed a greater adsorptive capacity than the modified PVP polyamide ultrafilter. There was no TNFα-inducing activity detectable in samples taken from the filtrate port. These results agree with the concept that maximalising hydrophobicity for ultrafilters will augment their margin of safety.

In summary, the polyamide membrane removes pyrogenic, cytokine-inducing substances derived from gram-negative bacteria contaminating aqueous solutions. Both molecular weight exclusion and adsorption are important mechanisms of removal. Ultrafiltration using hydrophobic polyamide ultrafilters provides a method with a wide margin of safety to produce pyrogen-free fluids for parenteral use in large quantities.

The Polyamide Dialyzer Membrane in in vitro Hemodialysis: No Evidence of Pyrogen Permeability under Conditions of Diffusive Transport across the Membrane

Compared to the hydrophobic parent polyamide membrane used for ultrafiltration, a more hydrophilic (PVP-modified) version of the polyamide is used for high-flux hemodialysis (Polyflux dialyzer 130, 160). In hemodialysis, diffusion is the predominant transport mechanism. Using highly permeable membranes in hemodialysis may favor the passage of endotoxin from dialysate into the blood [18]. This is of particular importance as bicarbonate dialysate of present standards has been shown to be highly contaminated with gram-negative water bacteria in approximately 10% of all dialysis centers studied [3]. This study also describes that *Pseudomonas* species are the predominant microorganisms found in contaminated dialysate. Besides the known pyrogenic cell-wall components (LPS, lipid A, muramyl peptides, etc.) *Pseudomonas aeruginosa* has recently been shown to produce an exotoxin as well [1]. Furthermore, the first evidence that *P. maltophilia,* a microorganism commonly found in dialysate, releases pyrogenic, peptide-like substances of low molecular weight (< 5 kD) which are very potent cytokine inducers but negative in the LAL assay has been presented [2]. It is presumed that these substances are biologically active fragments of a *P. maltophilia* exotoxin yet to be identified, purified and sequenced.

It has previously been demonstrated that under the conditions of in vitro dialysis with diffusion being the predominant transport mechanism across the dialyzer membrane, low-flux regenerated cellulosic membranes are permeable to IL-1-inducing endotoxin-like substances derived from cultured *E. coli* [19]. Further studies were performed to test the permeabil-

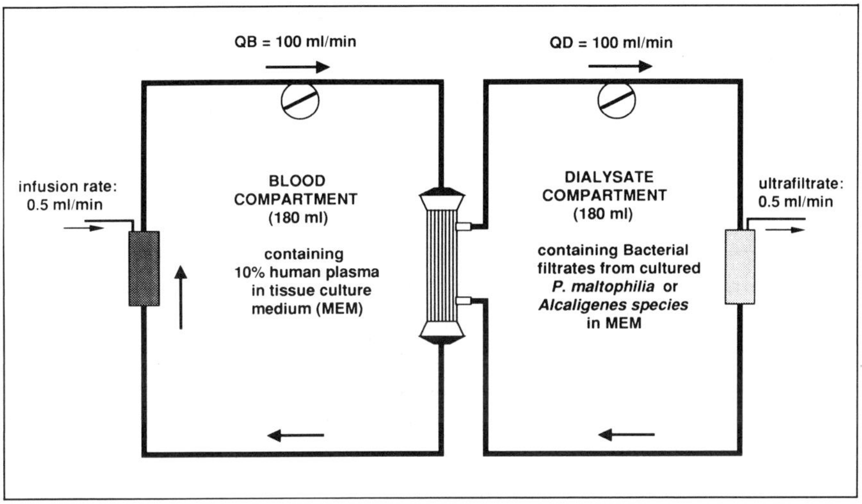

Fig. 9. The closed loop in vitro dialysis circuit. Two of these systems were run in parallel, dialyzers were kept at 37 °C. Compartment volumes, flow rates and the ultrafiltration rate are indicated. Circulation was performed for 2 h in a countercurrent manner.

ity of polyamide dialyzers (Polyflux) to TNFα inducing substances derived from dialysate-borne bacteria in comparison to regenerated cellulose [20].

The in vitro Dialysis Circuit

A closed loop dialysis system as shown in figure 9 was used. The blood lines were standard PVC hemodialysis line sets. Regenerated cellulose (surface area: 1.2 m^2; membrane thickness: 8 μm) and Polyflux 130 (hydrophilic polyamide) (surface area: 1.3 m^2; membrane thickness: 50 μm) dialyzers were perfused in separate circuits in parallel experiments. The dialysis system was assembled under sterile conditions and the blood and dialysate compartments were rinsed with 2 liters of sterile, pyrogen-free saline. The saline on the dialysate side was then replaced by ultrafiltered Eagle's minimal essential medium (MEM), the volume approximately 180 ml, and circulated at a flow rate of Q_D = 100 ml/min. After 1 h of sterile dialysis, 18 ml of bacterial culture filtrate derived from *P. maltophilia,* a gram-negative organism grown from dialysate, were added to the dialysate compartment.

P. maltophilia was grown in brain heart infusion broth. Bacteria-containing broth was then centrifuged and the supernatant was filtered (pore size 0.22 μm) to obtain the culture filtrate which contained approximately 2 μg/ml endotoxin as measured by the LAL assay. The 1:10 dilution of bacterial filtrate in MEM (LAL activity approximately 0.2 μm/ml) was circulated for one additional hour. The blood compartment (180 ml) contained 10% human plasma (heparinized with 10 U/ml) in MEM. In all experiments the solution on the blood side was circulated at a flow rate of Q_B = 100 ml/min. Both loops were kept at 37 °C and circulated in a countercurrent manner during the 2 h of in vitro dialysis. On the blood side a constant infusion of MEM containing 10% plasma at a flow rate of 0.5 ml/min was maintained. As a counterpart, a syringe was hooked to the dialysate circuit to take up the ultrafiltered volume (0.5 ml/min). According to pressure measurements, which were obtained simultaneously at the blood inlet and outlet as well as at the dialysate inlet and outlet, the calculated backfiltration was zero for regenerated cellulose and 1.1 ml/min for polyamide in the closed loop system [20].

Incubation of Mononuclear Cells to Measure TNFα-Inducing Activity

Samples from the blood and the dialysate loop of the in vitro systems were drawn at 30 and 60 min (sterile dialysis) and 90 and 120 min (bacterial filtrate in dialysate) of recirculation. All samples were incubated with human MNC to measure TNFα-inducing activity. MNC were separated from donor blood and incubated as previously described [19]. Following incubation for 18 h, TNFα was measured in MNC supernatants by radioimmunoassay [17]. The TNFα-inducing activity is expressed as ng/2.5 $\times$ 10^6 MNC/18 h.

The known intra- and interindividual variability in the MNC response to bacterial substances (e.g. LPS) [17] was taken into account in the experimental design of this study: The same donor's plasma and cells separated from one blood sample were used to test the permeability of the two dialyzers in parallel in vitro circuits on the same day.

No Evidence for Pyrogen Permeability of the Polyamide Dialyzer Membrane during in vitro Dialysis

The results of five parallel experiments (means ± SEM) testing the permeability of regenerated cellulose and polyamide to TNFα-inducing substances derived from cultured *P. maltophilia* are shown in figure 10. Compared to the first 60 min of sterile dialysis addition of *P. maltophilia*

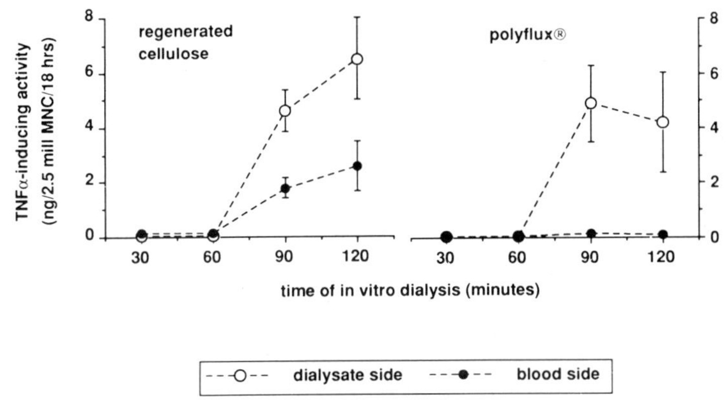

Fig. 10. TNFα production of human MNC incubated with samples from the dialysate compartment containing MEM (open circles) and from the blood compartment containing 10% plasma in MEM (closed circles), before (30 and 60 min) and after challenge (90 and 120 min) of the dialysate with *P. maltophilia* filtrate. Results for regenerated cellulose are depicted on the left panel, those for polyamide on the right. The time of in vitro dialysis is indicated under the horizontal axis. TNFα-inducing activity is expressed as ng/2.5 $\times$ 10^6 MNC/18 h. Symbols indicate the mean of 5 parallel experiments ± SEM. Derived from Lonnemann et al. [20].

filtrate increased the TNFα-inducing activity in the dialysate compartment within 30 min with regenerated cellulose to 21.4 ± 1.9 (ng/2.5 $\times$ 10^6 MNC/18 h) and with polyamide to 13.1 ± 2.8 (ng/2.5 $\times$ 10^6 MNC/18 h). In the presence of such high amounts of TNFα-inducing bacterial substances in the dialysate, blood side activity increased with regenerated cellulose to 3.7 ± 1.2 (ng/2.5 $\times$ 10^6 MNC/18 h), $p < 0.01$, but was unchanged compared to sterile in vitro dialysis when polyamide dialyzers were used.

Recent work has confirmed this data that LAL-negative MNC-positive permeation of intact dialyzer membranes occurs after challenge by LPS-derived material [21]. Thus, only the incubation of material with human MNC provides a test system which detects all relevant biologically active substances derived from water bacteria.

Conclusions

The hydrophilic polyamide dialysis membrane was tested under conditions of diffusive transport across the membrane, which may facilitate the passage of small pyrogens. Like the hydrophobic polyamide ultrafiltration membrane, this membrane also seems to be a barrier rejecting cytokine-inducing substances derived from water bacteria with a wide margin of safety. Effective adsorption seems to be based on hydrophobic interactions between the membrane surface and hydrophobic domains in the pyrogenic material. Evidence is presented favoring maximal hydrophobicity when the polyamide membrane is used as an ultrafilter. The mechanism of adsorption of pyrogens by polyamide is uncertain but seems to depend in part upon the availability of domains of hydrophobicity.

References

1 Liu PV: Exotoxins of *Pseudomonas aeruginosa* factors that influence the production of exotoxin Am J Infect Dis 1973;116:481–489.
2 Mahiout A, Lonnemann G, Schulze M, Shaldon S, Koch KM: Mononuclear cells release cytokines after stimulation with LAL-negative small (< 1 kD) bacterial products (abstract). Nephrol Dial Transplant 1990;5:706–707.
3 Klein E, Pass T, Harding GB, Wright R, Million C: Microbial and endotoxin contamination in water and dialysate in the central United States. Artif Organs 1990;14:85–94.
4 Dinarello CA: Interleukin-1 and interleukin-1 antagonism. Blood 1991;77:1627–1652.
5 Duff GW, Atkins E: The detection of endotoxin by in vitro production of endogenous pyrogen: Comparison with *Limulus* amebocyte lysate gelation. J Immunol Methods 1982;52:323–331.
6 Dinarello CA, Krueger JM: Induction of interleukin-1 by synthetic and naturally occurring muramyl peptides. Fed Proc 1986;45:2545–2548.
7 Takahashi I, Kotani S, Takada H, Shiba T, Kusumoto S: Structural requirements of endotoxic lipopolysaccharides and bacterial cell walls in induction of interleukin-1. Blood Purif 1988;6:188–206.
8 Dinarello CA, O'Connor JV, LoPreste G, Swift RL: Human leukocytic pyrogen test to detect pyrogenic material in growth hormone from recombinant *E. coli.* J Clin Microbiol 1984;20:323–329.
9 Matthiasson E: The role of macromolecular adsorption in fouling of ultrafiltration membranes. J Membr Sci 1983;16:23–36.
10 Reihanian H, Robertson CR, Michaels AS: Mechanisms of polarization and fouling of ultrafiltration membranes by proteins. J Membr Sci 1983;16:237–258.

11 Margenau H, Kestner NR: Theory of Intermolecular Forces. New York, Pergamon Press, 1969.
12 Israelachvili JN: Intermolecular and Surface Forces with Applications to Colloidal and Biological Systems. New York, Academic Press, 1985.
13 Norde W: Adsorption of proteins from solution at solid-liquid interface. Adv Colloid Interface Sci 1987;25:267–340.
14 Dinarello CA, Lonnemann G, Maxwell R, Shaldon S: Ultrafiltration to reject human interleukin-1-inducing substances derived from bacterial cultures. J Clin Microbiol 1987;25:1233–1238.
15 Gery I, Waksman BH: Potentiation of the T-lymphocyte response to mitogens. II. The cellular source of potentiating mediator(s). J Exp Med 1972;136:143–155.
16 Kaye J, Gillis S, Mizel SB, Shevach EM, Malek TR, Dinarello CA, Lachman LB, Janeway CA Jr: Growth of a cloned helper T-cell line induced by a monoclonal antibody specific for the antigen receptor: Interleukin-1 is required for the expression of receptors for interleukin-2. J Immunol 1984;133:1339–1346.
17 Van der Meer JWM, Endres S, Lonnemann G, Cannon JG, Ikejima T, Okusawa S, Gelfand JA, Dinarello CA: Concentrations of immunoreactive tumor necrosis factor alpha produced by human mononuclear cells in vitro. J Leuk Biol 1988;43:216–223.
18 Hosoya N, Sakai K: Backdiffusion rather than backfiltration enhances endotoxin flux through highly permeable dialysis membranes. Trans Am Soc Artif Intern Organs 1990;36:311–313.
19 Lonnemann G, Bingel M, Floege J, Koch KM, Shaldon S, Dinarello CA: Detection of endotoxin-like interleukin-1-inducing activity during in vitro dialysis. Kidney Int 1988;33:29–35.
20 Lonnemann G, Behme TC, Lenzner B, Floege J, Mahiout A, Colton CK, Koch KM, Shaldon S: The permeability of dialyzer membranes to TNFα-inducing substances derived from water bacteria. Submitted.
21 Laude-Sharp M, Caroff M, Simard L, Pusineri C, Kazatchkine MD, Haeffner-Cavaillon N: Induction of IL-1 during hemodialysis: Transmembrane passage of intact endotoxins (LPS). Kidney Int 1990;38:1089–1094.

Dr. G. Lonnemann, Department of Nephrology,
Medizinische Hochschule Hannover, D-W-3000 Hannover (FRG)

Therapeutic Applications

Shaldon S, Koch KM (eds): Polyamide – The Evolution of a Synthetic Membrane for Renal Therapy. Contrib Nephrol. Basel, Karger, 1992, vol 96, pp 64–76

Hemofiltration in the Past and Present

K. Schaefer, D. von Herrath

St. Joseph Hospital, Berlin, FRG

Chronic hemofiltration is a clinically interesting, effective and safe blood-purification method which has been in use for 15 years for the treatment of end-stage renal disease (ESRD) [1]. It is intriguing that although this method offers many theoretical advantages compared to conventional hemodialysis, hemofiltration has never become generally accepted. Recent data show that the total number of patients treated regularly with chronic hemofiltration has remained constant over the past 5 years, about 2,500. The proportion of patients on hemofiltration has, however, gone down, since the total dialysis population has increased. The share of European hemofiltration patients is today around 2.5% [2].

In 1991, if one poses the question, why hemofiltration developed as an alternative therapy to hemodialysis, the answer must include the widespread dissatisfaction with standard hemodialysis (acetate/cuprophan) in the mid-1970s. During the late 1970s and early 1980s, a large percentage of patients became hypotensive and suffered from cardiovascular problems under dialysis. Various investigators found that hemofiltration treatment involved fewer hemodynamic problems. Hemofiltration also seemed particularly interesting because the membrane used for this method permitted removal of substances that low-flux hemodialysis could not eliminate or only in small quantities [3].

This review will present several technical developments of hemofiltration and in addition clinical aspects that in the past have been of interest to nephrologists. The present state of the art with certain advantages of hemofiltration as a chronic renal therapy will then be presented.

Some Technical Aspects of Hemofiltration

Hemofiltration requires continuous infusion of a substitution solution. The first balancing system for chronic hemofiltration was introduced in 1973 [4, 5]. In contrast to the technique commonly used today, which is characterised by introduction of the substitution solution after the hemofilter (postdilution mode), the pioneering work of Henderson was with the predilution system. Quellhorst et al. [1] developed the first gravimetric system which made it possible to infuse substitution solution after the filter. This postdilution mode made the clinical application of hemofiltration feasible as the volume of reinfusion fluid required was significantly reduced.

Why then are only a relatively small number of European patients treated with postdilution hemofiltration? The answer is that hemofiltration in comparison to hemodialysis has a significantly higher treatment cost. The volume of substitution solution that must be infused into the patient corresponds to at least 30–40% of the body weight. Since this solution must be of intravenous quality, it represents a considerable cost. Hemofiltration machines that produce substitution fluid in-line (see below) are a recent development which has reduced cost [6]. Hemofiltration can now be carried out with automated machines. The blood side is identical to a standard hemodialysis machine. The fluid balancing system is connected to an electronic scale with a computer-controlled infusion pump. The treatment dose, i.e. the desired exchange volume, is selected and automatically achieved. The infusion rate is controlled by the filtration capacity of the hemofilter. The weight loss is linear and spread over the entire treatment period. Hemofilters with a polyamide membrane are particularly good for hemofiltration, since they make filtration rates of over 180 ml/min possible, although this level is only reached in cases where there is an adequate blood flow. As a rule, the blood flow should not fall below 500 ml/min. The substitution liquid, which is administered continuously, may vary in its composition, especially regarding sodium and calcium. Sodium concentration is usually between 135 and 145 mmol/l, calcium between 1.9 and 2.1 mmol/l. It is important that the electrolyte concentrations are adjusted so that no loss of important electrolytes occurs.

A further important technical aspect of hemofiltration is the sterility required of the substitution liquid. Bacterial contamination may occur during the production of the solution into the bag (usually containing 4–5 liters). Contamination may also occur through small leaks in the infusion bags during storage. Once this problem was recognised, a small hollow fiber ultrafilter (polyamide, cut-off 40,000 Da) was placed between the infusion bag and the patient to reduce the risk of bacterial and pyrogen contamination of the stored solution. The danger of bacterial contamination has been reduced by the introduction of in-line production of infusion solution. This is characterised by production of the solution from treated tap water and dialysate concentrate. Bacteria and pyrogens are removed by the use of particle filters, a charcoal filter, a water softener, a reverse osmosis unit and finally various polyamide ultrafilters. Repeated studies have always shown that infusion solutions prepared in this way are sterile and pyrogen free. Patients receiving in-line hemofiltration suffer fewer fever reactions than patients on hemodialysis [6]. Recently, a new in-line hemofiltration system has been developed (Gambro, GHS 10, Lund, Sweden) in which the substitution solution is prepared from high-quality RO-water and concentrate. Before it enters the blood, it must pass through either a final bacteria filter as well as an ultrafilter fitted with a polyamide membrane, or a double ultrafilter. As mentioned above, an electronic flow measuring system regulating the addition of substitution liquid in accordance with the filtrate guarantees a linear weight loss. We reported recently on our clinical experience with this new in-line hemofiltration system [6]. In over 1,300 therapies (45,500 liters of substitution liquid produced) we observed no side effects. The errors in weight control between a gravimetric balance and the flow control system showed a maximal difference of 110 g/treatment. We observed no instances of bacterial or endotoxin contamination in the substitution liquid during these treatments and no febrile reactions were observed [6]. This positive experience has continued in subsequent years. It can be stated that substitution liquid produced automatically (in-line) by the technique outlined above represents a significant step forward in establishing chronic hemofiltration. It must, however, be admitted that, although the cost as compared to conventional hemofiltration has been reduced, it is still significantly higher than for hemodialysis.

Clinical Results of Chronic Hemofiltration

In view of the fact that only a few publications on this subject have appeared recently, we would like to describe in the following several clinical results achieved with chronic hemofiltration. We would also like to

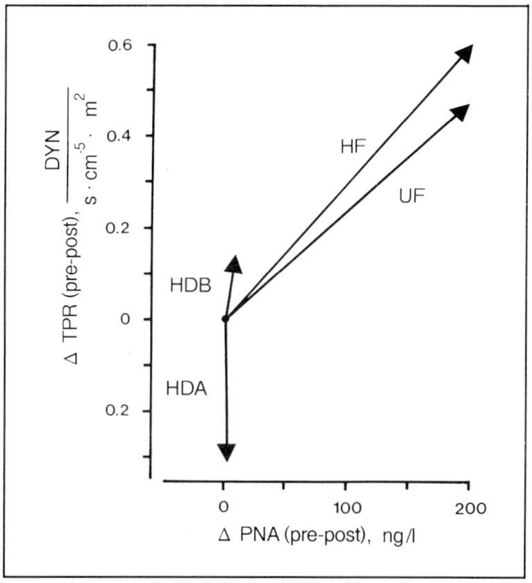

Fig. 1. Comparison of hemofiltration (HF), isolated ultrafiltration (UF), bicarbonate hemodialysis (HDB) and acetate hemodialysis (HDA) with regard to sympathetic response. TPR = Total peripheral resistance; PNA = plasma noradrenaline. Derived from Baldamus et al. [8].

mention a recent publication which reviews the important technical and clinical advantages and outcome parameters associated with hemofiltration [7]. Finally, we will discuss the influence of hemofiltration on the removal of β_2-microglobulin (β_2-M).

Cardiovascular Stability

Blood pressure drops occur in about 20–30% of all cases of conventional hemodialysis (cuprophane/acetate). Shortly after the introduction of conventional postdilution hemofiltration in the treatment of ESRD, it was demonstrated by various authors that hemofiltration has a particularly positive effect on the cardiovascular system in hypotensive as well as hypertensive patients [3]. A final explanation of the mechanism whereby hemofiltration leads to better vascular stability remains a matter of speculation. It is of interest that an increase in peripheral vascular resistance (PVR) has been demonstrated during isolated ultrafiltration and hemofiltration, whilst PVR has remained stable with bicarbonate dialysis and often decreases during acetate dialysis (fig. 1) [8, 9].

One of the most exciting theories advanced to explain the superiority of hemofiltration in this respect is the interleukin hypothesis [10, 11]. It postulates that exposing whole blood to dialysis membranes and contaminated dialysate results in the activation of peripheral blood mononuclear cells (PBMC) to produce pro-inflammatory cytokines. The consequences of this could lead to chronic inflammatory lesions seen in long-term dialysis patients such as muscle wasting, fibrosis, premature ageing and dialysis-related amyloidosis. If sufficiently activated PBMC can result in acute febrile reactions, hypotension and even shock.

Hemodialysis could induce the production of pro-inflammatory cytokines by different mechanisms. Stimulation of the monocytes on the blood side of the dialysis membrane could occur when C5a binds to the cell surface. Alternatively, dialysate-derived pyrogenic materials could adhere to the membrane on the dialysate side and subsequently pass through the membrane by the mechanism of charge and flow dynamics. These pyrogens would then induce the PBMC to produce interleukin-1 and TNFα. The latter explanation seems to be the more likely, as mere adsorption of monocytes to the membrane also occurs during isolated ultrafiltration with cuprophane membranes where hypotensive events are seldom observed. One can therefore speculate that the flow of dialysate is in fact a crucial condition for the production of pro-inflammatory cytokines. The diffusion or backfiltration from dialysate to blood of solutes and contaminants with the capacity to induce cytokine synthesis is probably the principal mechanism for monocyte stimulation. It is probable that such events will be less likely to occur during isolated ultrafiltration or hemofiltration.

Although the mechanism for the superiority of hemofiltration remains unexplained, a recent publication by Quellhorst compared 10,000 hemodialysis and hemofiltration treatments retrospectively with regard to the frequency of hypotensive reactions [12]. He concluded that the incidence of hypotension increases in a linear relation to the amount of liquid removed in hemodialysis. At the same time, hemofiltration allowed a reduction of the initial body weight by 2% without causing significant blood pressure reduction. We also observed this positive influence of hemofiltration on circulatory stability [6]. The positive effect of hemofiltration on cardiovascular stability was recently demonstrated in a prospective study [12]. Patients who had received hemodialysis for 12 weeks and then changed over to hemofiltration showed a clear improvement in cardiovascular stability. It was concluded that the higher extracellular sodium concentration in hemofiltration contributed to its cardiovascular stability.

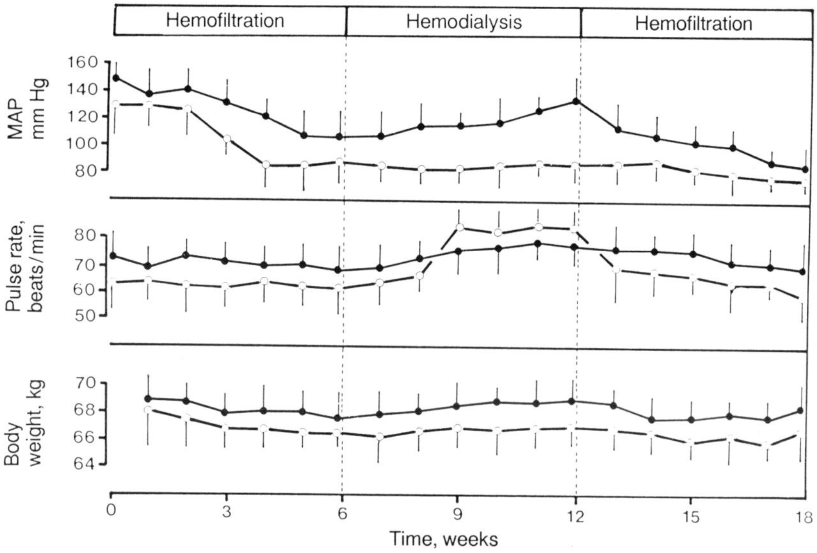

Fig. 2. Change of blood pressure, pulse rate and body weight during a cross-over study beginning with chronic hemofiltration, followed by hemodialysis and then by hemofiltration for a second period. ● = Pre-treatment value, ○ = post-treatment value. Derived from Quellhorst et al. [14].

Hypertension

The beneficial influence of hemofiltration in patients with severe hypertension was claimed many years ago [13, 14]. Indeed, hypertension was an indication for hemofiltration, as it was observed that the blood pressure could be controlled within months of starting hemofiltration [13, 14] (fig. 2). An explanation for this phenomenon was that hemofiltration was efficient in controlling volume-dependent hypertension because the dry body weight was more easily achieved than with hemodialysis [15–17].

Various Uremic Complications

Shortly after chronic hemofiltration became generally available many reports were published claiming that hemofiltration could be considered as the panacea for a large number of uremic complications [14–20]. It was demonstrated that hemofiltration was capable of ameliorating the acid-base status in patients switched from conventional acetate hemodialysis

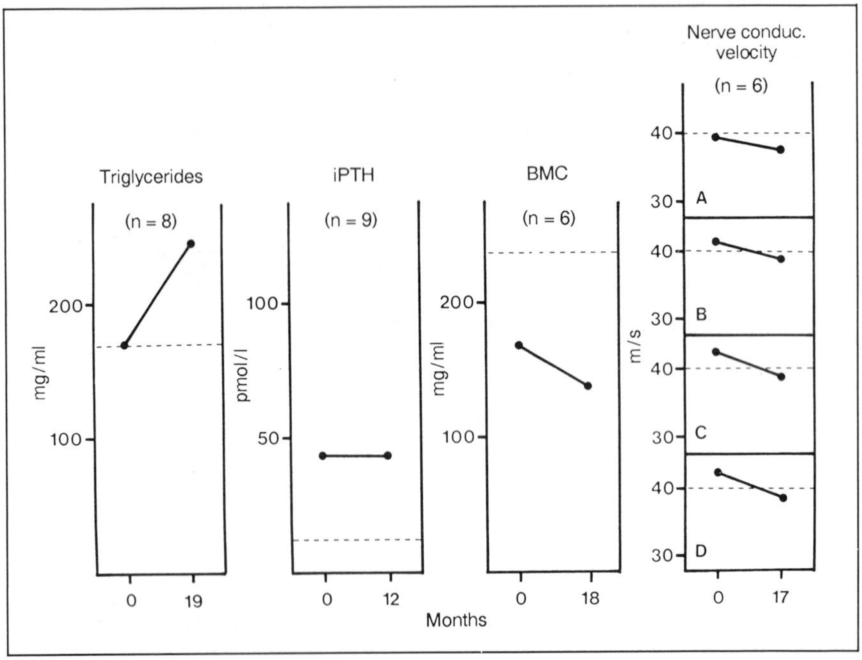

Fig. 3. Influence of chronic HF on various uremic complications. BMC = Bone mineral content. Nerve conduction velocity measured in arms (A + B) and legs (C + D). Derived from Schaefer et al. [30].

[21, 22]. The postulated superiority with regard to various uremic complications was based on the fact that hemofiltration could remove medium-sized and larger molecules. This initial optimism was derived from reports claiming that some uremic complications are in part caused by larger molecules, such as parathyroid hormone (PTH) [23, 24]. It was shown that intact PTH as well as PTH fragments can be removed during hemofiltration [25, 26]. Regarding serum calcium, it became evident that unlike hemodialysis, hemofiltration has almost no impact on the ionized serum calcium, one mechanism by which hemodialysis exerts an influence on PTH secretion [27]. A critical evaluation of the impact of hemofiltration on various so-called uremic complications is still difficult to undertake, since this method has only been practised on small patient populations and for a few years. Thus, although it is not possible to draw final conclusions at present it has to be stated that in this respect the available data do not

reveal any superiority for hemofiltration over hemodialysis [28–32]. As summarized in figure 3, neither serum triglycerides, bone mineral content nor nerve conduction velocity in arms and legs showed any significant change after periods of prolonged hemofiltration treatment for months to years. However, there appears to be some recent evidence that patients maintained for several years on hemofiltration present a serum PTH concentration which is lower than that of patients on hemodialysis for a comparable time.

Clinical Outcome

The influence of chronic hemofiltration on the survival rate of patients with ESRD as compared to chronic hemodialysis merits comment. No difference in survival between the two techniques was found in ages between 18 and 60 for uncomplicated ESRD patients from one center [33]. However, in patients over 60 years of age, those treated with hemodialysis had an actuarial 5-year survival rate of 73% compared to 80% with hemofiltration. In a more recent joint study with Quellhorst, we were able to demonstrate an actuarial survival rate for 62 patients, with primary renal causes of ESRD, over 60 years of age at the start of treatment, which was significantly better than a comparable group of hemodialysis control patients with uncomplicated ESRD (see fig. 5) [34]. These data suggest that hemofiltration may be the method of choice for the older patient with ESRD who is not eligible for transplantation.

β_2-Microglobulin

It is now accepted that a major component of the amyloid associated with chronic hemodialysis is a new form of amyloid fibril protein which is derived from β_2-microglobulin (β_2-M) [35]. Recent results clearly demonstrated that hemofiltration using polyamide hemofilters removed significantly more β_2-M than cuprophan hemodialysis (fig. 4) [36, 37]. The serum concentration of β_2-M was reduced by more than 70%, with a negative mass balance of almost 300 mg of β_2-M during a single hemofiltration treatment with hydrophilic polyamide membrane. The serum β_2-M concentration in a patient with β_2-M amyloidosis was 45 mg/l prior to commencement of hemofiltration therapy and fell to below 30 mg/l after 55 weeks of hemofiltration [38]. Post-treatment serum concentration, after 55 weeks, were at almost normal values, between 6 and 8 mg/l (fig. 5). Although it is unclear whether this reduction of serum β_2-M has any influence on the regression or possible prevention of β_2-M amyloidosis, it

4
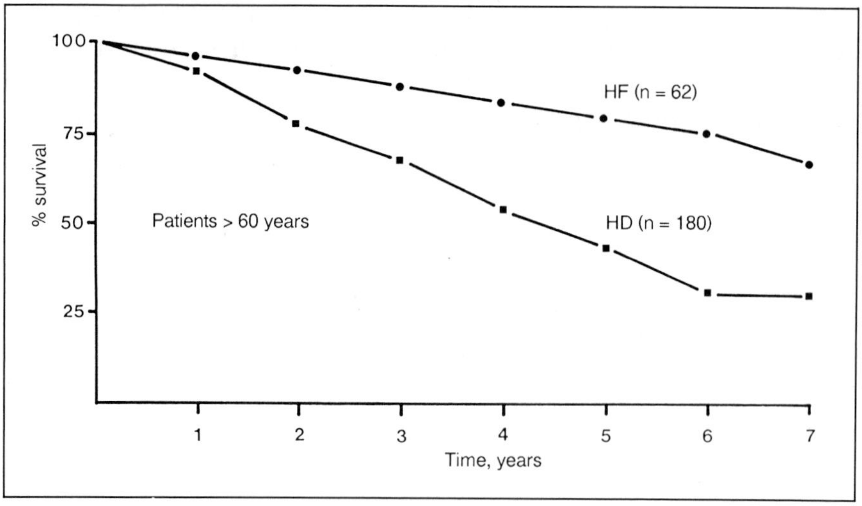

5
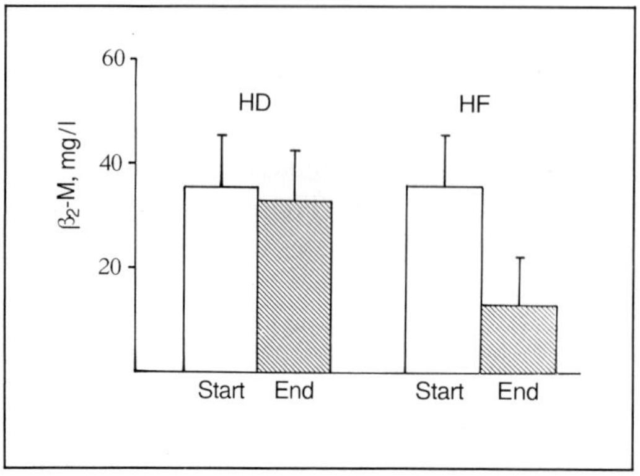

Fig. 4. Survival of elderly uremic patients treated either with HF or HD. Derived from Schaefer et al. [34].

Fig. 5. Concentration of β_2-M in blood before and after treatment with cuprophan hemodialysis (HD) or polyamide hemofiltration (HF).

would seem reasonable to attempt to lower the precursor amyloidogenic protein to as low a level as possible. It is obvious because of the molecular weight of β_2-M (11,800 Da) that convective transport will offer the most efficient removal method even compared to high-flux dialysis. Postdilutional hemofiltration relies exclusively on this transport method for its treatment of ESRD. This assumption is supported by a recent suggestion

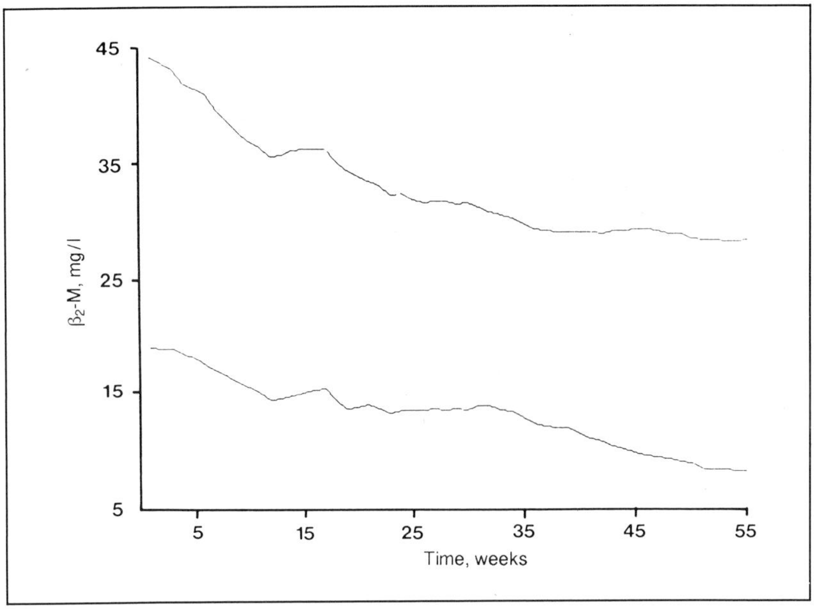

Fig. 6. Pre- and post-treatment β_2-M serum levels in a patient treated for 55 weeks with hemofiltration.

that there was a lower incidence of carpal tunnel syndrome in hemofiltration patients compared to hemodialysis patients treated for a similar period of time [39].

The suggested lower incidence of dialysis-related amyloidosis in hemofiltration patients may not be due solely to the improved removal β_2-M. As was initially stated sterile substitution liquid is used in hemofiltration. The use of sterile dialysate has, on the other hand, only recently been suggested and is not yet a routine procedure [40, 41]. As discussed above, a contaminated dialysate can induce production of pro-inflammatory cytokines, which may then augment the synthesis of β_2-M. It is of particular interest that a recent report using sterile, pyrogen-free dialysate with cuprophan dialyzers claimed a reduction in the patients' β_2-M serum concentration [39]. Further work is needed to verify these preliminary claims on the beneficial effect of using a sterile dialysate with cuprophan dialyzers, as it is now possible with a polyamide ultrafilter to produce sterile dialysate relatively easily.

Conclusions

Chronic hemofiltration can be considered an important life-sustaining therapy in the treatment of ESRD. Previous assumptions regarding the superiority of hemofiltration in influencing several classic uremic complications can today no longer be sustained. Neither lipid disorders, nor polyneuropathies nor anemia were influenced more favorably using this method than under conventional cuprophane dialysis. The longer survival times in older patients on chronic hemofiltration are of importance as it is just this segment of the ESRD population that is not going to receive a transplant.

In addition, the potential to produce in-line sterile, pyrogen-free infusion fluid with bicarbonate for hemofiltration should add benefits to the pronounced vascular stability already seen with acetate hemofiltration.

Although, hemodialysis in the 1990s will be characterised by three components: (1) dialysate containing bicarbonate, (2) high-flux biocompatible membranes, and (3) sterile pyrogen-free dialysate, the role of hemofiltration in the treatment of the geriatric ESRD patient is currently assured.

References

1 Quellhorst E, Rieger J, Doht B, Beckmann H, Jacob I, Kraft B, Mietzsch G, Scheler F: Treatment of chronic uraemia by an ultrafiltration kidney. First clinical experiences. Proc Eur Dial Transplant Assoc 1976;13:314–321.

2 Fassbinder W, Brunner FP, Bryuger H, Ehrich JHH, Geerlings W, Raine AEG, Rizzoni G, Selwood NH, Tufveson G, Wing AJ: Combined report on regular dialysis and transplantation in Europe, XX, 1989. Nephrol Dial Transplant 1991;6(suppl I): 5–35.

3 Schaefer K, Koch K-M, Quellhorst E, von Herrath D: Hemofiltration. Contr Nephrol. Basel, Karger, 1982, vol 32, pp 1–223.

4 Henderson LW, Besarb A, Michaels A, Bluemle LW: Blood purification by ultrafiltration and fluid replacement (Diafiltration). Trans Am Soc Artif Intern Organs 1967;16:216–222.

5 Henderson LW, Livoti LG, Ford CA, Kelly AB, Lysaght MJ: Clinical experience with intermittent hemofiltration. Trans Am Soc Artif Intern Organs 1973;19:119–123.

6 Schaefer K, von Herrath D, Hartenstein-Koch K, Kutschera D: Clinical experience with a new in-line hemofiltration system: An analysis of 1300 treatments. Int J Artif Organs 1989;12(suppl 4):5–8.

7 Henderson LW, Quellhorst EA, Baldamus CA, Lysaght MJ (eds): Hemofiltration. Berlin, Springer, 1986.

8 Baldamus CA, Ernst W, Fassbinder E, Koch KM: Differing haemodynamic stability due to differing sympathetic response: Comparison of ultrafiltration, haemodialysis and hemofiltration. Proc Eur Dial Transplant Assoc 1980;17:205–212.

9 Baldamus CA, Ernst W, Kachel HG, Lysaght M, Koch KM: Hemodynamics in hemofiltration. Contr Nephrol. Basel, Karger, 1982, vol 32, pp 56–60.

10 Henderson LW, Koch KM, Dinarello CA, Shaldon S: Hemodialysis hypotension: The interleukin hypothesis. Blood Purif 1983;3–8.

11 Shaldon S, Deschodt G, Branger B, Granolleras C, Baldamus CA, Koch KM, Lysaght M, Dinarello CA: Haemodialysis hypotension: The interleukin hypothesis restated. Proc Eur Dial Transplant Assoc 1985;22:229–243.

12 Quellhorst E: Cardiovascular problems in end-stage renal failure hemofiltration. Contr Nephrol. Basel, Karger, 1990, vol 78, pp 212–220.

13 Quellhorst E, Schuenemann B, Doht B: Treatment of severe hypertension in chronic renal failure by haemofiltration. Proc Eur Dial Transplant Assoc 1977;14:129–135.

14 Quellhorst E, Schuenemann B, Hildebrand U, Neumann W: Hypertension and haemofiltration. Contr Nephrol. Basel, Karger, 1982, vol 32, pp 46–55.

15 Schaefer K, von Herrath D, Gullberg CA, Asmus G, Hüfler M, Offermann G, Cremer H, Heuck CC, Ritz E: Chronic hemofiltration. A critical evaluation of a new method for the treatment of blood. Artif Organs 1978;2:386–394.

16 Baldamus CA, Schoeppe W, Koch KM: Comparison of haemodialysis (HD) and post dilution haemofiltration (HF) on an unselected dialysis population. Proc Eur Dial Transplant Assoc 1978;16:228–235.

17 Schneider H, Streicher E, Hövelborn U, Müller HAG, Spohr U, Schmidt-Gayk H: Haemofiltration – critical evaluation of clinical benefits. Proc Eur Dial Transplant Assoc 1979;16:218–224.

18 Pierides AM, Giacherio D, Schniepp B, Burritt M: Calcium, phosphorus and parathyroid hormone metabolism in chronic hemofiltration. Contr Nephrol. Basel, Karger, 1982, vol 32, pp 77–85.

19 DeFremont JF,Moriniere P, Pruna A, Gheerbrandt JD, Kassouf J, Roussel A, Galy C, Fournier A: Long-term evaluation of hemofiltration. Contr Nephrol. Basel, Karger, 1982, vol 32, pp 119–127.

20 Schneider H: Electrolyte balance during hemofiltration treatment. Contr Nephrol. Basel, Karger, 1982, vol 32, pp 111–118.

21 Bosch JP, Saccaggi A, Glabman S: Alternatives in extracorporeal uremia therapy; in Nissenson AR, Fine RN, Gentile DE (eds): Clinical Dialysis. pp 681–704, New York, Appleton-Century, 1984.

22 Schaefer K, Ryzlewicz T, Sandri M, von Bernewitz S, von Herrath D: Effect of hemofiltration on acid-base status and ventilation. Contr Nephrol. Basel, Karger, 1982, vol 32, pp 69–78.

23 Massry SG: Is parathyroid hormone a uremic toxin? Nephron 1978;19:125–130.

24 Massry SG: The toxic effects of parathyroid hormone in uremia. Semin Nephrol 1983;4:306–328.

25 Schaefer K, Offermann G, von Herrath D, Asmus G, Hüfler M: Parathyroid hormone, 25-OH-vitamin D, and digoxin levels in patients treated by chronic hemofiltration. J Dial 1977;1:169–630.

26 Schaefer K, Offermann G, Asmus G, Hüfler M: Verhalten von Calcium, Phosphat und Parathormon unter chronischer Hämofiltration. Nieren-Hochdruckkrankh 1978;1:40–43.
27 Schaefer K, von Herrath D, Offermann G: Long-term experiences with chronic hemofiltration. J Artif Organs 1980;3:219–220.
28 Schaefer K, von Herrath D, Hüfler M: Hemofiltration 1982: A stocktaking. Contemp Dial 1982;8:14–18.
29 Schneider H, Streicher E: Klinische Erfahrungen mit der intermittierenden Hämofiltrationsbehandlung. Nieren-Hochdruckkrankh 1978;1:21–28.
30 Schaefer K, von Herrath D, Asmus G, Offermann G: Is chronic hemofiltration better than chronic hemodialysis? Artif Organs 1979;3:40–43.
31 Henning HV, Balusek E: Lipid metabolism in uremia: Effect of regular hemofiltration and hemodialysis. J Dial 1977;1:595–605.
32 Fuchs C, Armstrong VW, Cremer P, Henning HV, Wieland H, Quellhorst E, Seidel A: An investigation of the lipoprotein profiles of patients on hemofiltration as compared to those on hemodialysis and intermittent peritoneal dialysis. Contr Nephrol. Basel, Karger, 1982, vol 32. pp 92–96.
33 Quellhorst EA, Schuenemann B, Mietzsch G: Long-term hemofiltration in 'poor-risk' patients. Trans Am Soc Artif Intern Organs 1987;33:758–764.
34 Schaefer K, Asmus G, Quellhorst E, Pauls A, von Herrath D, Jahnke J: Optimum dialysis treatment for patients over 60 years with primary renal disease. Survival data and clinical results from 242 patients treated either by haemodialysis or hemofiltration. Proc Eur Dial Transplant Assoc 1984;21:510–523.
35 Gejyo F, Odan S, Yamaoa T, Honma M, Saito H, Suzuki Y, Nakagawa Y, Kobayashi H, Marayuma Y, Hirasawa Y, Suzuki M, Arakawa M: β_2-Microglobulin. A new form of amyloid protein associated with chronic hemodialysis. Kidney Int 1986;30: 385–390.
36 Kaiser JP, Hagemann J, von Herrath D, Schaefer K: Different handling of β_2-microglobulin during hemodialysis and hemofiltration. Nephron 1988;48:132–135.
37 Schaefer K, Kaiser JP, von Herrath D, Göhl H: The striking effet of a new polyamide-hemofilter on the serum β_2-microglobulin in hemodialysis patients (abstract). Kidney Int 1988;33:237.
38 Schaefer K, Kaiser JP, Gohl H, Hagemann J, von Herrath D: Impressive reduction of serum β_2-microglobulin in a patient treated with a new polyamide hemofilter. Am J Nephrol 1990;10:173–174.
39 Quellhorst E, Schuenemann B: β_2-Microglobulin and hemofiltration; in Gejyo F, Brancaccio D, Bardin T (eds): Dialysis Amyloidosis, pp 123–129. Milano, Wichtig Editore, 1989.
40 Erley CM, von Herrath D, Hartenstein-Koch K, Kutschera D, AmirMoazami B, Schaefer K: Easy production of sterile, pyrogen-free dialysate. Trans Am Soc Artif Intern Organs 1988;34:205–207.
41 Erley CM, von Herrath D, Amir-Moazami B, Schaefer K: Beneficial effect of a bacteria and pyrogen reduced dialysate on the occurrence of fever during hemodialysis. A prospective study. J Nephrol 1989;4:257–269.

Prof. K. Schaefer, Medical Ward II, St. Joseph Hospital, Bäumerplan 24,
D-W-1000 Berlin 42 (FRG)

Shaldon S, Koch KM (eds): Polyamide – The Evolution of a Synthetic Membrane for Renal Therapy. Contrib Nephrol. Basel, Karger, 1992, vol 96, pp 77–85

Hemofiltration: Predilution versus Postdilution

S. David, V. Cambi

Cattedra di Nefrologia, Parma, Italia

Hemofiltration is characterized by a more efficient removal of high-molecular-weight solutes and a lower incidence of cardiovascular symptoms, in comparison to hemodialysis [1].

Vascular Stability

The factors responsible for better vascular stability and asymptomatic weight loss during hemofiltration are not yet completely understood. It has been demonstrated that convection allows a more physiologic hemodynamic adaptation to ultrafiltration, since it brings about an increase of vascular peripheral resistance which effectively offsets hypovolemia [2]. This mechanism of adaptation is blunted in hemodialysis. Furthermore, it has been demonstrated that a less negative sodium balance in hemofiltration reduces plasma osmolality changes, which helps prevent hypovolemia [3]. The genesis of hemodialysis-induced vascular instability is generally considered multifactorial, and several causes have been identified [4]. Due to technical progress most of them can be avoided today, without having to use convection. Bicarbonate substituting for acetate as the dialysate buffer, high sodium in the dialysate in order to minimize osmotic changes, weight loss rate tailored to the patient's tolerance, ultrapure dialysis fluid and biocompatible dialysis membranes; all these measures improve the vascular stability. The question whether convective transport may add further benefits is still unanswered.

Solute Transport

The transport of solutes in hemofiltration depends on the water flux through the membrane and on the solute rejection by the membrane. Membrane rejection is responsible for the lower solute concentration in the ultrafiltrate in comparison to plasma, and this difference is large for high-molecular-weight solutes. Solute rejection is influenced by several variables which prevent it from being constant. A deeper insight into this is useful for a better understanding of transport by convection and for its rational application.

Solute transport in hemofiltration as well as in isolated ultrafiltration is never purely convective. It is always a mixed convective-diffusive transport, because the ultrafiltered fluid can be compared to dialysis fluid. Paradoxically, the contribution of the diffusive transport is higher for high-molecular-weight solutes; however, with increasing ultrafiltration the diffusive component is progressively reduced. Because of the mixed transport the sieving coefficient is not constant but depends on the conditions. At low ultrafiltration rates diffusive solute transport contributes to a high sieving coefficient. With increasing ultrafiltration there is less diffusion and the sieving coefficient is reduced. When the ultrafiltration rate is high, the solutes rejected by the membrane accumulate on its surface. This solute polarization creates a layer where the concentration is higher than in the bulk phase plasma. As a consequence, the water flux through the membrane will drag a larger quantity of solute than predictable from the plasma concentration. The result is an increased solute concentration in the ultrafiltrate and an increased sieving coefficient.

Henderson et al. [5], utilizing dextrans of different molecular weight added to plasma, confirmed that the sieving coefficient is effectively reduced when the ultrafiltration rate increases. However, in the presence of very high ultrafiltration rates, the sieving coefficient did not increase again, as could be anticipated on the basis of solute polarization. His conclusion was that solute polarization in the course of hemofiltration is not an important phenomenon. Furthermore, he found no reduction of the sieving coefficient when the plasma flow was increased up to 700 ml/min.

This observation is in contrast with the theoretical assumption that with increased blood flow, the wall shear rate also increases, which reduces the phenomenon of solute polarization. This in turn should give a lower sieving coefficient.

Effect of Ultrafiltration Rate on Sieving

In order to verify these theoretical assumptions in vivo, we selected six chronic dialysis patients with high output arteriovenous fistulae. All patients underwent postdilution hemofiltration performed according to two different procedures: (1) Ultrafiltration rate varying from 30 to 150 ml/min and constant blood flow at 400 ml/min. (2) Constant ultrafiltration rate at 90 ml/min and blood flow varying from 300 to 600 ml/min.

Mean Hct was 25 ± 5% and plasma protein 6.6 ± 6 g/dl. Hydrophilic polyamide hemofilters (FH 88H, Gambro) were used. Ninety minutes after the beginning of the treatment, allowing sufficient time for maximum adsorption of $\beta_2$microglobulin (β_2m) by the membrane, the ultrafiltration rate or the blood flow was randomly changed within the limits of the experiment. After a stabilization period, the β_2m sieving coefficient was estimated from the β_2m concentration in the ultrafiltrate and in the plasma water. In order to exclude membrane adsorption the mass balance error was calculated, and when it was higher than 50% the test was considered not valid.

The changes of the sieving coefficient for β_2m under conditions of constant blood flow and changing ultrafiltration rate are shown in figure 1. When the ultrafiltration rate increases, there is a decrease of the sieving coefficient. At constant ultrafiltration rate the sieving coefficient increases with the blood flow up to values of 450–550 ml/min, but decreases again when the blood flow rate is very high (fig. 2).

The results are in agreement with the theoretical assumption that the sieving coefficient is dependent on both ultrafiltration and on solute polarization, but are partially in contrast with the in vitro tests reported in the literature. It is possible that in vivo the GR effect, as shown by Colton et al. [6] promotes a more efficient mixing of the blood unstirred fluid film on the membrane and reduces, at very high blood flow, the solute polarization phenomenon. In postdilution hemofiltration, an increased blood flow promotes increased ultrafiltration and favors small solute clearance. However, the clearance of high-molecular-weight solutes, such as β_2m, is not always

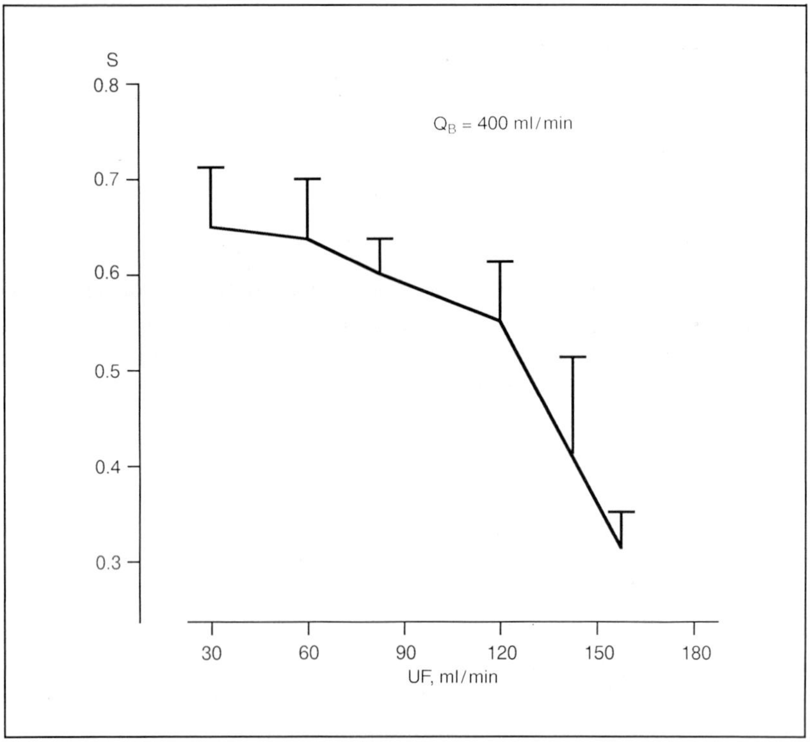

Fig. 1. Variations in sieving coefficient for β_2m at different ultrafiltration rates and constant blood flow (400 ml/min). In vivo conditions. Mean values ± SD.

improved by a higher blood flow, since the effect of increased ultrafiltration is partially blunted by the reduction of the sieving coefficient. The final result may be that for similar Kt/V urea values, the β_2m removal does not increase when a very high blood flow has been employed to reduce the treatment time.

Solute Removal in Predilution Hemofiltration

In postdilution hemofiltration the ultrafiltration increases proportionally with TMP up to a plateau, where further TMP increases are ineffective [7]. This is due to the hemoconcentration and solute polarization within the hemofilter, which increase the oncotic pressure thus offsetting the

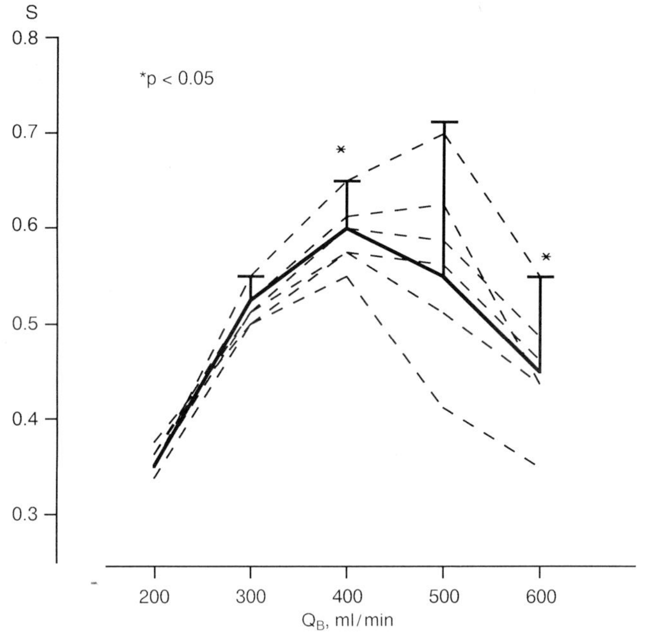

Fig. 2. Variations in sieving coefficient for β_2m at different blood flow rates and constant ultrafiltration (90 ml/min). In vivo conditions. Mean values ± SD.

transmembrane hydraulic pressure. These effects are magnified by further hemoconcentration due to the progressive weight loss during the treatment. It has been suggested that the filtration fraction should not exceed the value of 0.5 [8] and as a consequence the plasma flow is the limiting factor for ultrafiltration. In order to obtain higher ultrafiltration rates, giving treatment times similar to standard hemodialysis (240 min), very high blood flow rates are necessary, and these may be difficult to obtain in practice [9]. With the predilution method it is possible to achieve an increased ultrafiltration rate [6], as the hemodilution reduces all the physical phenomena which oppose the production of ultrafiltrate from blood. Figure 3 shows the viscosity in plasma and blood within a dialyzer in the course of post- and predilution hemofiltration, respectively.

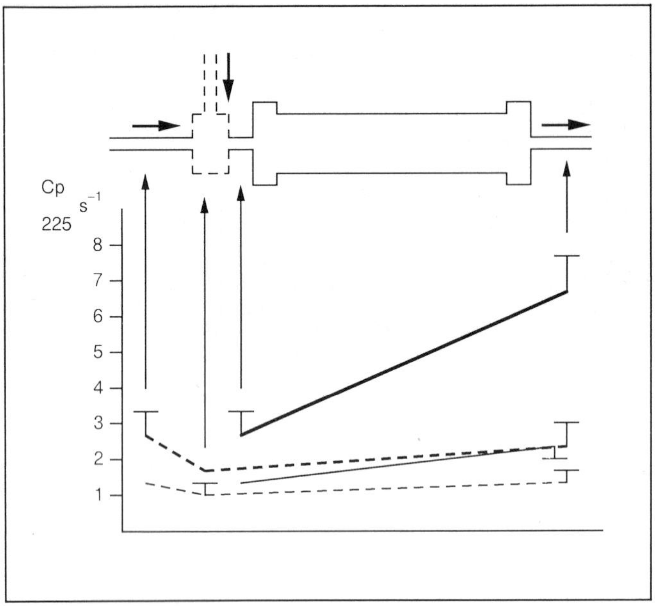

Fig. 3. Viscosity changes of blood (narrow) and plasma (bold) in postdilution (unbroken) and predilution (dashed) hemofiltration. The arrows indicate the sampling points. Postdilution: Q_B = 500 ml/min, UF = 190 ml/min; predilution: Q_B = 400 ml/min, UF = 406 ml/min.

We compared predilution hemofiltration to postdilution hemofiltration in the same population and established that the treatment efficiency was clearly increased during predilution (table 1). In these studies hemofilters with hydrophobic polyamide were used, so the results cannot be directly compared with our data above. In predilution it is possible to obtain urea clearance of about 190 ml/min with a blood flow of 400 ml/min. In order to obtain similar results in postdilution hemofiltration blood flows higher than 550–600 ml/min are required. Furthermore, in predilution the ultrafiltration rate is more stable and the reduction of the urea clearance less pronounced (fig. 4).

As far as the high-molecular-weight clearance is concerned, it can be anticipated that predilution, thanks to the reduced hyperviscosity within the hemofilter, improves the sieving coefficient compared to postdilution. The change of β_2m sieving coefficient in both pre- and postdilution is shown in figure 4. It can be observed that there are no significant differ-

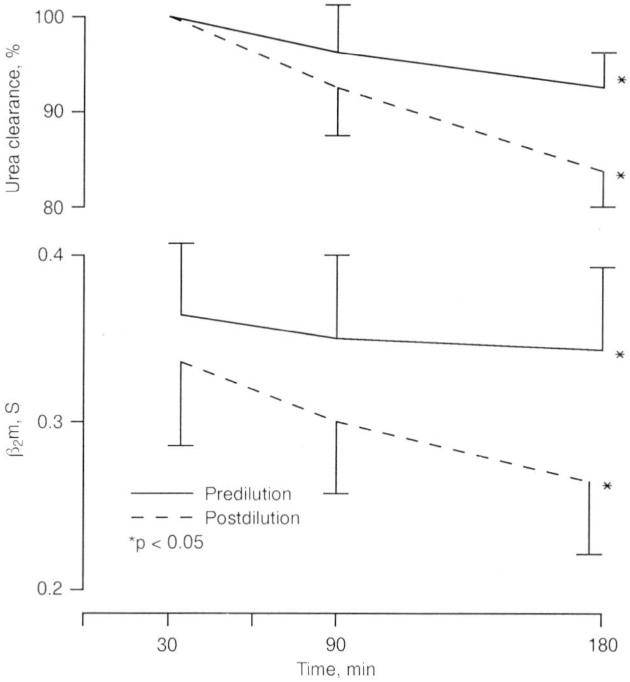

Fig. 4. Stability of urea clearance and β_2m sieving coefficient during pre- and postdilution hemofiltration. For treatment conditions see table 1.

Table 1. Treatment conditions during pre- and postdilution hemofiltration with 8 patients using 2 m^2 hydrophobic polyamide hemofilters (FH 88, Gambro)

	Postdilution hemofiltration	Predilution hemofiltration
Blood flow, ml/min	535 ± 42	391 ± 22
Substitution volume, liters	35 ± 3.5	69.1 ± 9.3
Ultrafiltration rate, ml/min	174 ± 35	385 ± 14
Urea clearance, ml/min	174 ± 35	191 ± 13
Kt/V	0.96 ± 0.04	0.97 ± 0.1
Treatment time, min	201 ± 17	180 ± 16

ences in the value of the sieving coefficient at the beginning of the treatment when pre- and postdilution are compared. However, after the second hour the sieving coefficient shows a significant drop in the postdilution mode. These results show that the sieving properties of the membrane are better preserved in predilution hemofiltration. In addition, the higher ultrafiltration rates make it possible to obtain a larger solute removal. The absence of significant differences in sieving coefficient in the initial phase of both modes of treatment does not confirm the hypothesized advantage of a reduced blood viscosity in predilution. On the other hand, it is possible that the increase of blood flow in predilution, which is produced by the real blood flow plus the volume of infusate (together 800 ml/min), increases the wall shear rate and reduces the solute polarization. This does not take place in postdilution with a blood flow of 450–500 ml/min.

Discussion

Solute removal by diffusion is selective to passage of substances, because of the physical resistance of the membrane and the blood film. As a consequence the removal of high molecular weight substances is inefficient in diffusive therapies, independently of their metabolic significance. Solute removal by convection, however, uniformly removes any solute irrespective of its molecular weight.

Treatment by convection can be applied using predilution of the blood as well as postdilution. Postdilution hemofiltration is limited by the increasing viscosity, the high TMP, and the progressively inefficient solute removal. This can be partially overcome by increasing the treatment time, while very high blood flows are necessary to improve the clearance of solutes. Predilution hemofiltration, with an infusion rate of 400–450 ml/min, can provide clerance of small solutes, which is competitive with hemodialysis within acceptable treatment durations (3–4 h). In addition, the removal of large solutes is superior to any other renal replacement therapy.

To date advances in hemodialysis have permitted a safer and more asymptomatic clinical treatment. Bicarbonate, controlled ultrafiltration, pyrogen-free dialysis fluid and more biocompatible membranes have significantly improved the clinical tolerance of hemodialysis. With the exception of bicarbonate, hemofiltration has comprised these factors from the start. The future will show whether predilution hemofiltration with bicarbonate instead of acetate will offer further clinical advantages in comparison to modern hemodialysis.

References

1 Quellhorst E, Rieger J, Doht B, Beckmann H, Jacob I, Kraft B, Mietzsch G, Scheler F: Treatment of chronic uremia by an ultrafiltration kidney – first clinical experience. Proc Eur Dial Transplant Assoc 1976;13:314–321.

2 Baldamus CA, Ernst W, Frei U, Koch KM: Sympathetic and hemodynamic response to volume removal during different forms of renal replacement therapy. Nephron 1982;31:324–332.

3 Gotch FA, Sargent JA: Hemofiltration: An unnecessarily complex method to achieve hypotonic sodium removal and controlled ultrafiltration. Blood Purif 1983; 1:9–15.

4 Baldamus CA, Pollok M: Ultrafiltration and hemofiltration: practical application; in Maher JF (ed): Replacement of Renal Function by Dialysis. Dordrecht, Kluwer Academic Publ, 1989, pp 327–346.

5 Henderson LH, Leypoldt JK, Frigon RP: The impact of membrane area on solute clearance in continuous arteriovenous hemofiltration; in La Greca G, Fabris A, Ronco C (eds): CAVH. Milan, Wichtig Editore, 1986, pp 37–47.

6 Colton CK, Henderson LW, Ford CA, Lysaght MJ: Kinetics of hemodiafiltration. I. In vitro transport characteristics of a hollow-fiber blood ultrafilter. J Lab Clin Med 1975;85:355–371.

7 Henderson LW: Biophysics of ultrafiltration and hemofiltration; in Maher JF (ed): Replacement of Renal Function by Dialysis. Dordrecht, Kluwer Academic Publ, 1989, pp 300–326.

8 Baldamus CA: Problems in hemofiltration. Contr Nephrol. Basel, Karger, 1985, vol 44, pp 212–222.

9 David S, Tagliavini D, Cambi V: Pre-post dilution hemofiltration. Nephrol Dial Transplant 1989;4:37–40.

Dr. S. David, Cattedra di Nefrologia, Via Gamsci 14, I–43100 Parma (Italy)

Shaldon S, Koch KM (eds): Polyamide – The Evolution of a Synthetic Membrane for Renal Therapy. Contrib Nephrol. Basel, Karger, 1992, vol 96, pp 86–98

Hemodiafiltration: Technical Aspects

Jan Sternby, Sven Jönsson, Ingrid Ledebo
Technical Research Department, Gambro AB, Lund, Sweden

The combination of convective and diffuse transport mechanisms in hemodiafiltration (HDF) offers the possibility to obtain an increased removal of both small and large solutes. Early clinical evaluations of this technique were made by Kunitomo et al. [1] and Leber et al. [2], but before that Shinaberger et al. [3] had used saline to limit the patient weight loss when testing membranes with high hydraulic permeability. The availability of highly permeable synthetic membranes, like polyamide, polysulfone and AN69, and of control systems designed specifically for use with these membranes, have made hemodiafiltration available for routine clinical practice. To optimize the treatment parameters a number of theoretical and practical issues need to be addressed. In this paper we will discuss theoretically how clearance is increased with increasing ultrafiltration rates, describe and evaluate various ways to perform hemodiafiltration, and finally study some of the limitations in ultrafiltration rates and pressures that apply to this form of therapy.

Clearance in Hemodiafiltration

For low-flux membranes it is easy to derive a formula for clearance under the assumption of zero ultrafiltration. It includes the dependence of clearance on dialyzer permeability and on blood and dialysate flows. With the use of more permeable membranes, as in hemodiafiltration, it becomes necessary to consider also the effect of ultrafiltration on clearance.

Calculating the ultrafiltration effect is quite complicated, and several formulae have been proposed. It is now well known that even for solutes with a sieving coefficient of one, simply adding the ultrafiltration rate to the diffusive clearance will overestimate the total clearance. The reason for this is that such an addition would assume that all ultrafiltrate were drawn from blood having the inlet concentration of the solute. In reality, the concentration of the ultrafiltrate will represent the average blood concentration over the length of the dialyzer.

A second effect of ultrafiltration is that it changes the flow rates, and thereby also the concentration profile along the dialyzer which affects the total diffusion. This is taken into account by Kunitomo et al. [1], van Geelen et al. [4] and Chang and Lee [5]. Furthermore, ultrafiltration will change the concentration profile across the membrane, which affects the local diffusion of solute. This effect was discussed by Sigdell [6], who gave a general clearance formula that involves numerical integration. As shown recently by Waniewski et al. [7], the case of unity sieving coefficient can be handled exactly. Clearance (K) for this case can be rewritten from their equation 20 as:

$$K = \frac{Q_B Q_D - f\,(Q_B - UF)\,(Q_D + UF)}{Q_D - f\,(Q_B - UF)}, \qquad (1)$$

where

$$f = \left[\frac{Q_B - UF}{Q_B}\;\frac{Q_D + UF}{Q_D}\right]^{1/\gamma} \text{ with } \gamma = e^{UF/kA} - 1,$$

Q_B and Q_D denote the blood and dialysate inlet flows, UF is the ultrafiltration rate and kA is the diffusive solute permeability of the dialyzer.

Different Hemodiafiltration Systems

The basic characteristic of hemodiafiltration in comparison to hemodialysis is increased convective transport through increased ultrafiltration. Various hemodiafiltration systems are used and differ mainly in the methods employed for the delivery of replacement fluid.

Hemodiafiltration with Preprepared Fluid in Bags (fig. 1a)

The early hemodiafiltration systems used preprepared sterile Ringer lactate solution [2]. For a given ultrafiltration rate the weight loss of the patient was gravimetrically controlled by manual adjustment of the infusion rate. Systems with automatic control of weight loss and infusion rate were later developed. The use of preprepared solutions in bags limits the choice of buffers in the substitution fluid to acetate or lactate due to precipitation of calcium carbonate in bicarbonate solutions [8]. *Paired filtration* is a special form of bag hemodiafiltration where the replacement fluid is infused between the hemofilter part and the hemodialyzer part of a special filter [9]. Thus, the convective and diffusive transports are separated from each other. In *biofiltration* the composition of the dialysis fluid and the substitution fluid is modified, so that bicarbonate is used only in the substitution fluid [10, 11].

Hemodiafiltration with Backfiltration (fig. 1b)

This form of controlled backfiltration is described in several papers [12–14]. The system contains two filters which play different roles. In the first filter as much ultrafiltrate as possible is taken out. In the second filter dialysate is reinjected into the bloodstream to give the desired patient weight loss. This is done by balancing the pressures with flow restrictors between the filters, either in the dialysate or in the blood stream or both. The large area of the two filters is not optimally utilized in this system, since backfiltration decreases the clearance in the second filter significantly.

Push-Pull Hemodiafiltration (fig. 1c)

By using the same filter sequentially in time for ultrafiltration and for backfiltration a system similar to backfiltration hemodiafiltration has been created. In *push-pull* hemodiafiltration the convective transport is increased during periods of extra high ultrafiltration, and the net weight loss is obtained by compensating with periods of backfiltration [15]. Clearance will then be increased during the high ultrafiltration periods and decreased during backfiltration, and the total push-pull clearance is a weighted mean of the two periods. Since the filter is only used for part of the time for ultrafiltration, the ultrafiltration rate has to be relatively higher to get the same efficiency.

On-Line Hemodiafiltration (fig. 1d)

A convenient way to produce large amounts of replacement fluid is by on-line filtration of the dialysis fluid [16]. For microbiological safety this must be combined with a strict disinfection routine [17]. Several clinical studies using the polyamide ultrafilter have shown that this method is both safe and feasible [18–20]. Freshly produced bicarbonate-containing re-

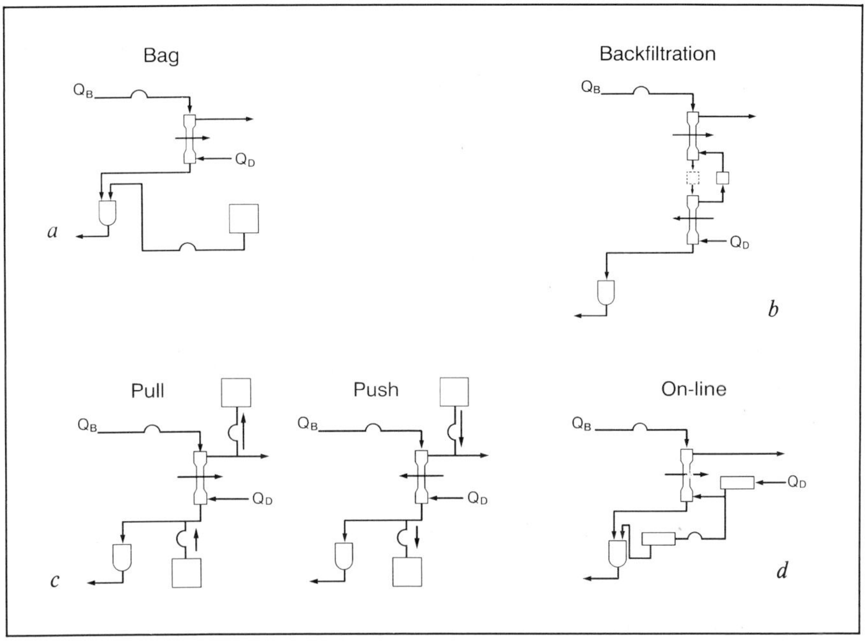

Fig. 1. Different ways of providing replacement fluid for hemodiafiltration (see text).

placement fluid of unlimited quantity can easily be used with the on-line system. This is a major advantage, since bicarbonate is known to improve the vascular stability [21].

Effects on Clearance

Clearance for two different size solutes as a function of ultrafiltration in both directions, i.e. also backfiltration, is illustrated in figure 2. During conditions of backfiltration only small solutes are cleared by diffusion. The clearance in push-pull hemodiafiltration is obtained by combining the values for high ultrafiltration and backfiltration. This may give an increased clearance for larger molecules, whereas small solute clearance is hardly affected. It is also obvious that normal hemodiafiltration at a constant, high ultrafiltration rate is always more effective than push-pull hemodiafiltration.

The diffusive clearance in on-line hemodiafiltration will be decreased by the reduction in dialysis fluid flow, due to the production of infusion fluid. It can, however, be shown that the total clearance always increases with increasing ultrafiltration. This is illustrated in figure 3. For larger solutes the effect of the reduced dialysate flow is negligible, and there is no difference between hemodiafiltration with bags and on-line hemodiafiltration.

Ultrafiltration during Hemodiafiltration Treatments: Limiting Factors

The optimal ultrafiltration rate during hemodiafiltration is a complex function of membrane type, area, blood flow and patient variables. It is common experience that the ultrafiltration rate achieved initially decreases during hemofiltration and hemodiafiltration treatments. To study some of the factors limiting the ultrafiltration rate during hemodiafiltration we performed the following studies.

Study 1 (in vivo):
Effect of Transmembrane Pressure and Time on Ultrafiltration Rate

The study consisted of three hemodiafiltration treatments performed on the same patient [22]. The membrane used was 1.6 m^2 polyamide (Polyflux 160), and the blood flow was 500 ml/min. During the three treatments of the same week the transmembrane pressure (TMP) was kept constant at either 150, 250 or 400 mm Hg. The infusion rate was controlled to achieve the desired weight loss.

Figure 4 shows that a higher ultrafiltration rate is obtained at higher TMP settings, but that there is a gradual decrease of ultrafiltration with time at all TMP values. The decrease is largest for the lowest TMP. This may be explained by a higher sensitivity at low TMP to the treatment-induced changes in blood composition [22]. The TMP influence also depends on the blood flow rate (fig. 5). At all blood flows the ultrafiltration rate increases with TMP up to a certain limit. At low blood flows, however, it may decrease with a further TMP increase [23].

One explanation for the limited ultrafiltration at high TMP values is that protein polarization at the membrane wall creates a decreased hydraulic permeability as well as an osmotic barrier [24]. The effect of this secondary membrane may increase during the treatment due to the reduction of the blood volume, which increases the hematocrit and the protein concentration. The blood cells could also have an effect on performance [25].

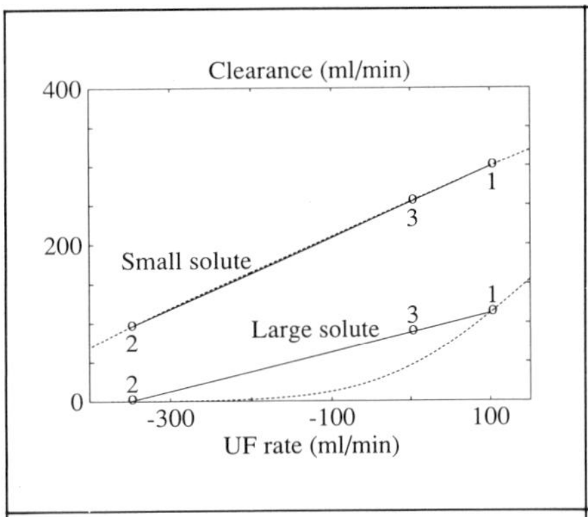

2

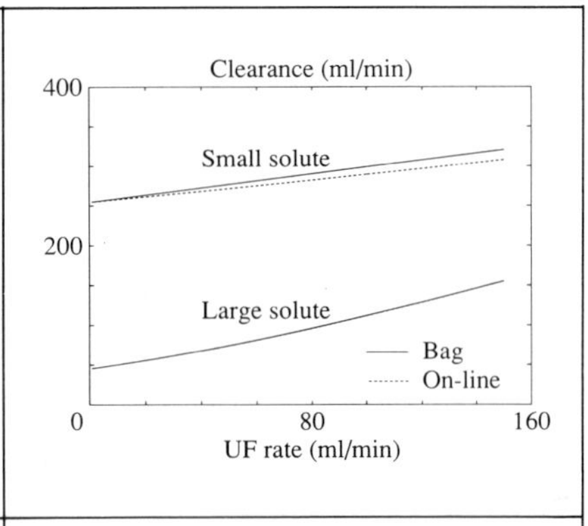

3

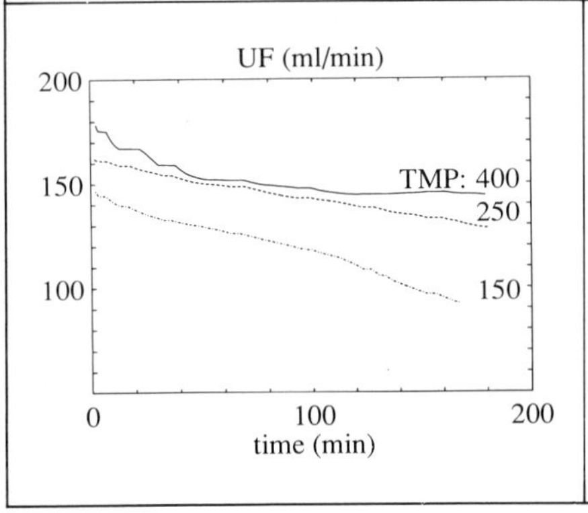

4

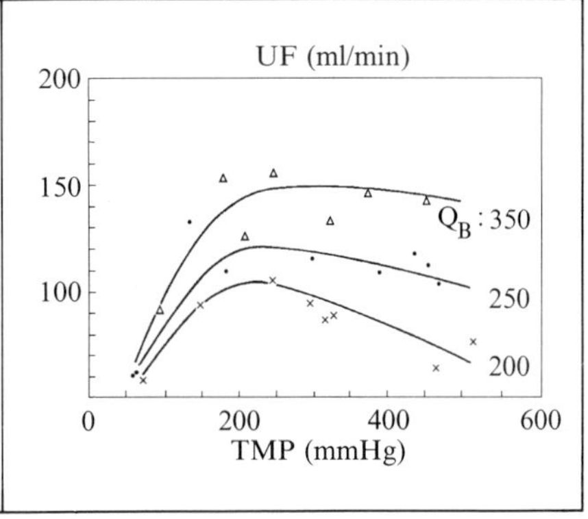

5

Fig. 2. Clearance as a function of UF in both directions (according to equation 1). The total clearance in push-pull HDF for a given weight loss (3) is found along the straight lines connecting the high UF (1) and the backfiltration UF (2). Q_B = 400 ml/min, Q_D = 500 ml/min and kA = 50 and 600 ml/min.

Fig. 3. Clearance as a function of UF in HDF with bags and in on-line HDF (according to equation 1). Q_D = 500 (HDF with bags) and Q_D = 500-UF (on-line HDF). Zero weight loss and Q_B and kA as in figure 2.

Fig. 4. UF rates achieved in vivo with Polyflux 160 during three HDF treatments at different TMP.

Fig. 5. Ex vivo UF rate versus TMP at different blood flows with a 0.9 m^2 polyamide flat sheet device. Bovine blood with hematocrit = 25%, and protein = 30.5 g/l. Data from [23].

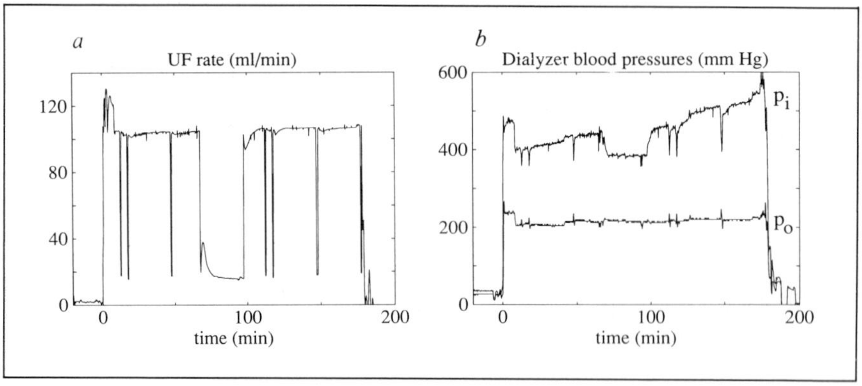

Fig. 6. (a) UF rate and *(b)* blood pressure at the inlet (p_i) and outlet (p_o) of the dialyzer during a HDF treatment with Polyflux 160.

Study 2 (in vivo):
Effect of Ultrafiltration Rate on the Pressure Drop over the Filter

The study consisted of one hemodiafiltration treatment performed on a patient regularly treated with hemodiafiltration. The membrane used was 1.6 m^2 polyamide (Polyflux 160) and the blood flow was 500 ml/min. The infusion flow was set manually, and the desired weight loss was achieved by automatic regulation of TMP. The ultrafiltration rate was thus directly dependent on the infusion rate, and is shown in figure 6a. The dialyzer inlet and outlet blood pressures are shown in figure 6b.

Two effects are worth noting. During the treatment there is a gradual increase of the inlet pressure, whereas the outlet pressure is constant. As a consequence, the pressure drop on the blood side of the dialyzer increases gradually. Secondly, this pressure drop depends on the ultrafiltration rate. When the ultrafiltration rate was decreased by decreasing the infusion rate to zero for a period just before minute 100, the pressure drop was minimized and remained constant.

To explain the dependence of the pressure drop on the ultrafiltration rate, we must consider the change in flow rate and blood viscosity within the dialyzer due to ultrafiltration. The blood viscosity depends on the hematocrit, which can be calculated from the hemoconcentration caused by ultrafiltration. For laminar flow, the local pressure drop per unit length is proportional to the product of viscosity and flow, which turns out to be fairly constant in the hematocrit range of 20–50%. Thus flow and viscosity changes

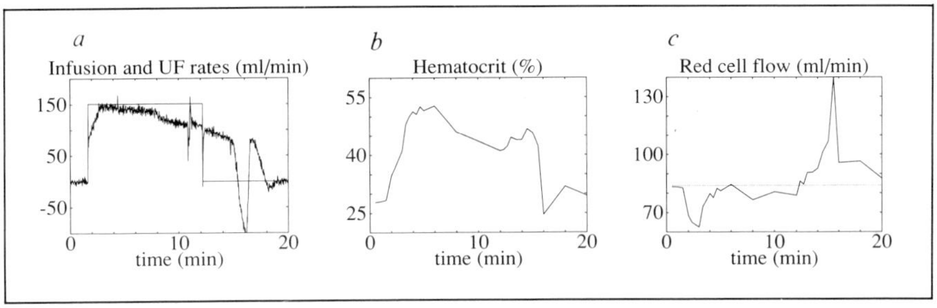

Fig. 7. Ex vivo circulation of bovine blood at 300 ml/min through a polysulfone filter (HF80) in HDF mode with zero weight loss. *a* Infusion and UF flow rates. *b* Hematocrit after the filter. *c* Red cell flow out of the filter (see text).

alone can not explain the observed pressure drop. The only factor left in the formula for laminar pressure drop is the fiber radius. This could be reduced if there is an adhesion of blood cells to the membrane wall. To explain the pressure changes observed the reduction would have to be in the order of 25% of the fiber radius. Such an effect has actually been observed and measured by several authors [26]. Studies with plasma give further evidence that the pressure drop is caused by blood cell deposition. The pressure drop with plasma is much lower than with blood and does not change with increasing ultrafiltration. In this case the flow decrease due to ultrafiltration is balanced by a viscosity increase due to protein concentration.

Study 3 (ex vivo):
Effect of Ultrafiltration Rate on Blood Cell Deposition

To study the adhesion of blood cells to the membrane we circulated bovine blood (hematocrit = 28%) through a 1.8 m^2 polysulfone device (HF80) at 300 ml/min under hemodiafiltration conditions with zero weight loss. Any set infusion rate would then be balanced by an equal amount of ultrafiltration through the filter, thus keeping the blood volume constant. The TMP required to give this ultrafiltration rate is automatically delivered by the machine. After an initial period of zero infusion and filtration, the infusion rate was increased to 150 ml/min, kept at this value for 10 min, and then returned to zero. Blood samples were taken every 20 s for hematocrit measurement.

Figure 7a shows the infusion and ultrafiltration flow rates. Ultrafiltration followed the infusion flow to start with, but required a steadily increasing TMP. When TMP reached the machine limit of 400 mm Hg, the ultrafiltration rate started to drop off. The measured hematocrit values after the dialyzer are displayed in figure 7b, and figure 7c shows the flow rate of red blood cells out of the dialyzer calculated from the measured flows and hematocrit. The ultrafiltration increase at 2 min caused a marked drop in red cell flow out of the dialyzer, and a corresponding peak was found as ultrafiltration decreased again at 15 min.

These changes are partly explained by hemoconcentration from the increased ultrafiltration, but are probably mainly due to cell adhesion to the membrane wall. The hematocrit peak which developed when TMP was decreased gives evidence of blood cell deposition. There was also an unexpectedly slow decrease of ultrafiltration when the infusion rate was decreased. This can be explained by a successively decreased hydraulic resistance as blood cells leave the membrane, which would sustain the ultrafiltration despite the decreased TMP. It is thus reasonable to believe that the ultrafiltration capacity of high-flux dialyzers is limited both by proteins and blood cells. The relative importance of these factors will depend on the operating conditions and on blood parameters.

Study 4 (ex vivo):
Effect of Blood Composition on Ultrafiltration Rate

Bovine blood (hematocrit = 26%, protein = 59.5 g/l) was circulated through a 2 m^2 polyamide hemofilter (FH88H) at 500 ml/min. The ultrafiltration flow across the membrane was increased in steps starting from zero by increasing TMP up to 400 mm Hg. The blood volume was kept constant by balancing the ultrafiltration with an equal amount of infusion. Thereafter, keeping TMP unchanged, the infusion rate was decreased to achieve hemoconcentration from an original blood volume of 10 l to 8. Finally, the blood volume was again kept constant.

Figure 8a shows the TMP schedule used in this experiment. The corresponding ultrafiltration rates achieved are shown in figure 8b, and the blood pressures at the inlet and outlet of the dialyzer are shown in figure 8c. Note that for each TMP level, the achieved ultrafiltration rate across the membrane and the blood pressures are constant as long as the blood volume is kept constant. As soon as the blood composition is changed by hemoconcentration, the ultrafiltration rate starts to decline and the inlet pressure increases. When the hemoconcentration is stopped, the ultrafiltration rate again becomes constant, but now at a lower level. The inlet pressure is also constant at a new higher level.

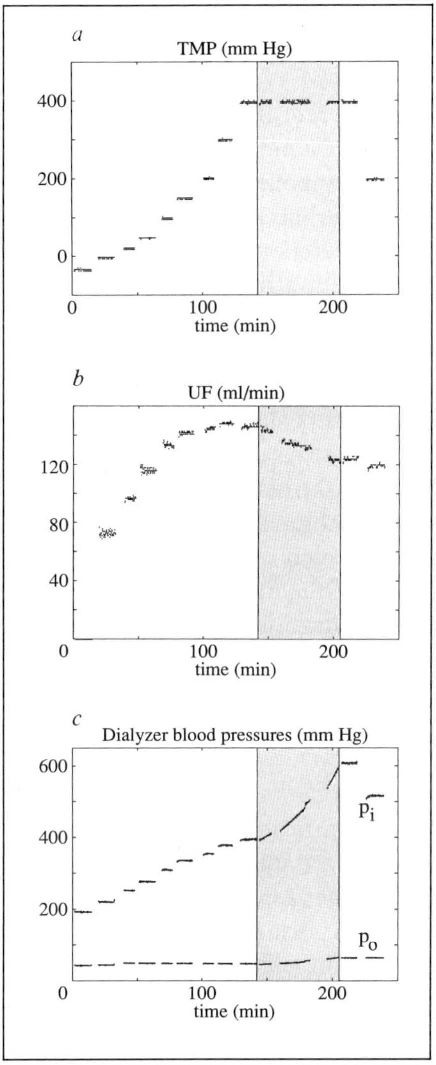

Fig. 8. Ex vivo circulation of bovine blood at 500 ml/min through a polyamide filter (FH88H). *(a)* TMP setting; *(b)* UF, and *(c)* blood pressure at the inlet (p_i) and outlet (p_o) of the dialyzer. The shaded area denotes a period of hemoconcentration. Values measured every 10 s during each period of constant TMP.

This experiment shows that the gradual changes of ultrafiltration rate and inlet pressure are mainly due to changes in the blood composition, and are not caused by intrinsic changes of the membrane itself. When selecting the TMP to achieve maximum ultrafiltration one must therefore take into account that the optimal TMP may change during the treatment as the composition of the blood is changing. This is especially evident at lower blood flows. With the high blood flow used in study 2 the maximum TMP was always optimal.

The limits to the maximum ultrafiltration rate also have an immediate consequence for the design of control systems for hemodiafiltration. When volumetric control is used, TMP will rise continuously, if the infusion rate is set above the ultrafiltration capacity of the dialyzer. Since the dialyzer capacity depends on several treatment parameters this may often happen in practice. The infusion rate thus needs to be controlled automatically by the machine.

To summarize these experimental studies, there is a limit to the ultrafiltration rate that can be achieved in hemodiafiltration treatments. The limit depends on the blood flow, the blood composition and on the dialyzer. It is not time dependent, but is affected by treatment-induced changes in blood composition. Another factor limiting the maximum ultrafiltration is the deposition of blood cells on the membrane. This also causes an increased pressure drop on the blood side, which may grow so large that very high pressures are created at the inlet of the dialyzer. In such a case the remedy would be to decrease the ultrafiltration rate rather than the blood flow, which might increase the pressure drop further.

Conclusions

As discussed in the theoretical part significant clearance improvements in hemodiafiltration, especially for larger solutes, can be achieved by increasing the ultrafiltration rate. This creates a need for larger volumes of replacement fluid. When using fluid in bags, cost and practical handling limit the volumes that can be used. With on-line preparation of fluid large volumes are easily provided and at a lower cost. Another important advantage with on-line preparation is the possibility to use bicarbonate as the buffer in the replacement fluid.

However, the high ultrafiltration rates also bring about other limitations. The permeability of the membrane will decrease due to the formation of layers of proteins and blood cells. Then ultrafiltration as a function of TMP will not increase above a certain maximum value. Also, the deposition of blood cells may give a high pressure drop in the dialyzer so that the inlet pressure becomes too high. This phenomenon could be further studied with a theoretical model.

Hemodiafiltration is a complex treatment form in theory as well as in practice. Theoretical models could therefore be of value to the clinicians when defining the optimum combination of treatment variables.

References

1 Kunitomo T, Lowrie EG, Kumazawa S, O'Brien M, Lazarus JM, Gottlieb MN, Merrill JP: Controlled ultrafiltration (UF) with hemodialysis (HD): Analysis of coupling between convective and diffusive mass transfer in a new HD-UF system. Trans Am Soc Artif Intern Organs 1977;23:234–243.

2 Leber HW, Wizemann V, Goubeaud G, Rawer P, Schütterle G: Simultaneous hemofiltration/hemodialysis: An effective alternative to hemofiltration and conventional hemodialysis in the treatment of uremic patients. Clin Nephrol 1978;9:115–121.

3 Shinaberger JH, Miller JH, Rubini ME, Gardner PW, Martin FE: Initial clinical evaluation of 'diafiltration'. Trans Am Soc Artif Intern Organs 1969;15:97–102.

4 van Geelen JA, Carpay W, Dekkers W, Fiers HA, Mulder AW, Flendrig JA: Simultaneous hemodialysis and hemofiltration: A simple, safe and effective treatment of uremic patients. Proc 2nd Symp Int Soc Artif Organs, 1979, pp 119–122.

5 Chang Y-L, Lee C-J: Solute transport characteristics in hemodiafiltration. J Membr Sci 1988;39:99–111.

6 Sigdell JE: Calculation of combined diffusive and convective mass transfer. Int J Artif Organs 1982;5:361–372.

7 Waniewski J, Werynski A, Ahrenholz P, Lucjanek P, Judycki W, Esther G: Theoretical basis and experimental verification of the impact of ultrafiltration on dialyzer clearance. Artif Organs 1991;15:70–77.

8 Feriani M, Biasioli S, Fabris A, Chiaramonte S, Ronco C, Bragantini L, Brendolan A, Dell'Aquila R, La Greca G: Calcium and bicarbonate containing solutions for peritoneal dialysis and hemofiltration; in Nose Y, Kjellstrand C, Ivanovich P (eds): Progress in Artificial Organs – 1985, pp 277–281. Cleveland, ISAO Press, 1986.

9 Ghezzi PM, Sanz-Moreno C, Gervasio R, Nigrelli S, Botella J: Technical requirements for rapid high-efficiency therapy in uremic patients. Paired filtration dialysis (PFD) with a two-chamber technique. Trans Am Soc Artif Intern Organs 1987;33: 546–550.

10 Zucchelli P, Santoro A, Raggiotto G, Esposti ED, Sturani A, Capecchi V: Biofiltration in uremia: Preliminary observations. Blood Purif 1984;2:187–195.

11 Bene B, Beruard M, Perrone B, Simon P: Simultaneous dialysis and filtration (SDF) with buffer-free dialysate (abstract). Blood Purif 1984;2:217.

12 Cheung AC, Kato Y, Leypoldt JK, Henderson LW: Hemodiafiltration using a hybrid membrane system for self-generation of diluting fluid. Trans Am Soc Artif Intern Organs 1982;28:61–65.

13 Shinzato T, Sezaki R, Usuda M, Maeda K, Ohbayashi S, Toyota T: Infusion-free hemodiafiltration: Simultaneous hemofiltration and dialysis with no need for infusion fluid. Artif Organs 1982;6:453–456.

14 Miller JH, von Albertini B, Gardner PW, Shinaberger JH: Technical aspects of high-flux hemodiafiltration for adequate short (under 2 hours) treatment. Trans Am Soc Artif Intern Organs 1984;30:377–381.
15 Usuda M, Shinzato T, Sezaki R, Kawanishi A, Maeda K, Kawaguchi S, Shibata M, Toyoda T, Asakura Y, Ohbayashi S: New simultaneous HF and HD with no infusion fluid. Trans Am Soc Artif Intern Organs 1982;28:24–27.
16 Henderson LW, Beans E: Successful production of sterile pyrogen-free electrolyte solution by ultrafiltration. Kidney Int 1978;14:522–525.
17 Mion CM, Canaud B: 'On-site' preparations of sterile apyrogenic electrolyte solutions for hemofiltration and hemodiafiltration; in Cambi V (ed): Short Dialysis, pp 261–291. Boston, Martinus Nijhoff, 1987.
18 de Lisle Nichols HF, Beau MC, Deschodt G, Branger B, Oules R, Mion C, Gullberg CA, Shaldon S: Experience with on-line hemofiltration (abstract). Blood Purif 1983; 1:49.
19 Canaud B, Imbert E, Kaaki M, Assounga A, Nguyen QV, Stec F, Garred LJ, Boström M, Mion C: Clinical and microbiological evaluation of a postdilutional hemofiltration system with in-line production of substitution fluid. Blood Purif 1990;8:160–170.
20 Rindi P, Pilone N, Ricco V, Cioni L: Clinical experience with a new hemodiafiltration (HDF) system. Trans Am Soc Artif Intern Organs 1988;34:765–768.
21 Biasioli S, Feriani M, Chiaramonte S, Cavallini L, Cesaro A, Fazion S, Petrosino L, Porena P, Zambello A: Different buffers for hemodiafiltration: A controlled study. Int J Artif Organs 1989;12:25–30.
22 Granolleras C, Jönsson S, Sternby J: Dialyzer ultrafiltration capacity during hemodiafiltration treatments (abstract). Blood Purif 1991;9:39.
23 Gullberg CÅ, Göhl H, Bergkvist G, von Herrath D, Schaefer K: A new hemofiltration system. Proc Eur Soc Artif Organs 1977;4:56–69.
24 Colton CK, Henderson L, Ford CA, Lysaght MJ: Kinetics of hemodiafiltration. I. In vitro transport characteristics of a hollow-fiber blood ultrafilter. J Lab Clin Med 1975;85:355–371.
25 Dorson WJ, Pierson JS: Controlling factors in hemofiltration. J Membr Sci 1989;44: 35–46.
26 Forstrom RJ, Bartelt K, Blackshear PL Jr, Wood T: Formed element deposition onto filtering walls. Trans Am Soc Artif Intern Organs 1975;21:602–608.

Dr. Jan Sternby, Gambro AB, Box 10101, S–220 10 Lund (Sweden)

Shaldon S, Koch KM (eds): Polyamide – The Evolution of a Synthetic Membrane for Renal Therapy. Contrib Nephrol. Basel, Karger, 1992, vol 96, pp 99–105

The Polyamide Membrane in Hemodiafiltration

M. Gonella[a], *G. Calabrese*[a], *G. Pratesi*[a], *A. Mazzotta*[a], *G. Vagelli*[a], *C. Baldin*[b]

[a] Servizio di Nefrologia e Dialisi, [b] Laboratorio Analisi, Ente Ospedaliero, Casale Monferrato, Italia

The use of highly permeable synthetic membranes in the treatment of chronic uremia is increasing. These membranes can be used either for pure convective transport, such as in hemofiltration, or for a combination of diffusion and convection, such as in hemodiafiltration (HDF) and high-flux dialysis. When utilized in HDF with the highest possible ultrafiltration the following goals are accomplished: (1) the risk of backfiltration of unsterile dialysate is avoided [1]; (2) the highest possible removal of small molecules is obtained [2]; (3) similar intradialytic vascular stability as in hemofiltration is achieved [3, 4]; (4) an efficient removal of higher molecular weight solutes, such as β_2-microglobulin, is obtained [5, 6].

The polyamide membrane which was originally developed for hemofiltration has subsequently been modified to make it more hydrophilic and enhance the diffusive permeability for small molecules. In dialyzers and hemodiafilters equipped with this membrane the fiber length is shortened and the internal diameter widened in order to provide optimal performance for these modes of therapy. Such devices equipped with this membrane (PA) are designated as Polyflux and are presently available with two different surface areas, 1.3 and 1.6 m^2. The purpose of the present study was to investigate the following aspects of Polyflux performance in HDF. (1) Ultrafiltration capacity and small solute clearance in vivo compared to other high-flux synthetic membranes. (2) Influence of the surface area on the clearance of urea, creatinine and phosphate at varying ultrafiltration rates. (3) Long-term results in patients undergoing HDF with Polyflux and comparison with the previous dialysis treatment.

Patients and Methods

Comparison between Polyflux and Other Synthetic High-Flux Dialyzers in HDF Mode

Dialyzers fitted with the following membranes were used in the same 5 patients. PA 1.6 m^2, AN69 1.6 m^2, PAN 1.8 m^2 and PMMA 2.0 m^2. In each patient, HDF was performed with two different techniques: (A) TMP set at 400 mm Hg in order to obtain the highest ultrafiltration rate; (B) spontaneous TMP (around 180 mm Hg) depending on blood flow (Q_b). In all cases Q_b was 400 ml/min, Q_d 500 ml/min; the substitution fluid (sterile bags) was infused in the postdilution mode; the session length was kept constant in the same patient during the eight studied sessions (four with each technique) [7]. In all the sessions, ultrafiltration rate (UF), urea and creatinine clearance (UrCl, CrCl) were measured.

Comparison between 1.3 m^2 and 1.6 m^2 Polyflux in HDF Mode

In another group of 4 patients, 1.3 m^2 and 1.6 m^2 polyamide dialyzers (Polyflux 130 and Polyflux 160) were employed in high ultrafiltration HDF (Q_b 400 ml/min, Q_d 500 ml/min, TMP 400 mm Hg) in postdilution mode. The session length was constant in each patient. In all the sessions, ultrafiltration rate, urea, creatinine and phosphate clearance (UrCl, CrCl, PiCl) were measured.

Comparison between HDF and HD Using Polyflux

In each of 3 patients, both 1.3 m^2 and 1.6 m^2 Polyflux dialyzers were used once with two different techniques: (A) postdilution high ultrafiltration HDF with Q_b 400 ml/min, Q_d 500 ml/min, TMP 400 mm Hg; (B) hemodialysis (HD) with Q_b 400 ml/min, Q_d 500 ml/min with volume control and linear weight loss using AK10 FCM (Fluid Control Monitor) to achieve desired body weight. The treatment times were kept constant for each patient in the two techniques. In all the sessions, ultrafiltration rate (UF) and urea, creatinine and phosphate clearances (UrCl, CrCl, PiCl) were measured.

Long-Term Evaluation of Patients on High UF HDF with Polyflux and Comparison with the Previous Treatment Mode

One group of 5 patients had been on high ultrafiltration HDF with 1.6 m^2 Polyflux for 13 ± 4 months and another group of 13 patients on high ultrafiltration HDF with 1.3 m^2 Polyflux for 22 ± 8 months. In both groups, dialysis time, post dialysis weight, predialysis urea and creatinine and postdialysis hemoglobin were compared. Out of the above 18 patients, 9 had been on the same mode of therapy with Polyflux since the beginning of their replacement therapy, whereas the other 9 started hemodialysis on a cuprophan dialyzer (Cu-HD) and were later moved to HDF with Polyflux (Polyflux-HDF). In this last group some parameters were compared at the beginning and the end of the period on Polyflux-HDF. Only 1 patient was on erythropoietin therapy during the HDF period.

Clearance and Laboratory Investigations

Only midweek sessions were considered in order to avoid large fluctuations of predialytic BUN and serum creatinine (sCr). Small toxin clearances were calculated with the following formula [8]:

$$K = \frac{C_d \times V_d \times \ln\left(\frac{C_i}{C_t}\right)}{T\left[C_i\left(\frac{C_i}{C_t}\right)^{\Delta t/T} - C_t\left(\frac{C_t}{C_i}\right)^{\Delta t/T}\right]},$$

where C_d represents the solute concentration in all the collected dialysate, V_d the volume of all dialysate. C_i and C_t are the plasma solute concentrations at 5 min from the beginning (t_1) and 5 min before the end (t_2) of the session, respectively; $\Delta t = t_2 - t_1$; T the time between C_i and C_t.

BUN, serum creatinine (sCr), serum phosphate (sPi), RBCs count and Hb level were determined by autoanalyzer methods.

Statistical Methods

The data were expressed as mean ± SD and compared by using the Student's t test for paired data.

Results

Comparison between Polyflux and Other Synthetic High-Flux Dialyzers

The values of UF and clearances obtained with the studied membranes are shown in table 1. As expected, UF with Polyflux (PA) was significantly higher ($p < 0.05$) with 400 mm Hg TMP HDF (technique A) than with spontaneous TMP (technique B). However, the difference in UF rate between the two techniques was less for Polyflux than for the dialyzers

Table 1. Mean values ± SD of the parameters examined in two techniques of HDF (A = TMP 400 mm Hg, B = spontaneous TMP) by using four high-permeability membranes: in all cases Q_b was 400 ml/min, Q_d 500 ml/min and the session length was 214 min

Parameters	PA 1.6m²		AN69 1.6 m²		PAN 1.8 m²		PMMA 2.0 m²	
HDF technique	A	B	A	B	A	B	A	B
TMP, mm Hg	400	180	400	180	400	180	400	180
UF, ml/min	121±11	106±5	119±12	79±14	119±14	83±10	113±8	72±8
Substituted volume liters	24±2	20±2	24±3	15±3	24±3	16±3	22±1	13±2
UrCl, ml/min	226±13	231±16	226±10	236±9	229±17	220±15	224±12	228±15
CrCl, ml/min	164±13	175±10	167±9	173±11	172±14	165±10	168±11	174±10

Table 2. Mean values ± SD of the parameters examined by using two different PA areas (1.3 and 1.6 m^2) in high UF HDF (Q_b 400 ml/min, Q_d 500 ml/min, TMP 400 mm Hg)

Parameters	PA 1.3 m^2	PA 1.6 m^2	p
Session length, min	201 ± 17	201 ± 17	
UF, ml/min	117 ± 6	131 ± 3	< 0.02
Substituted volume, liters	21 ± 3	24 ± 4	< 0.02
UrCl, ml/min	195 ± 11	239 ± 12	< 0.02
CrCl, ml/min	153 ± 4	175 ± 2	< 0.001
PiCl, ml/min	122 ± 9	150 ± 8	< 0.02

Table 3. Mean values ± SD of the parameters examined by using the two different PA areas in HD and HDF: in all cases Q_b was 400 ml/min and Q_d 500 ml/min

Technique	HD	HDF	HD	HDF
PA area, m^2	1.3	1.3	1.6	1.6
Session length, min	223 ± 27	223 ± 27	190 ± 18	190 ± 18
UF, ml/min	14 ± 3	113 ± 6	10 ± 3	133 ± 4
Substituted volume, liters	0	22 ± 1	0	24 ± 1
UrCl, ml/min	181 ± 9	209 ± 6	187 ± 7	244 ± 17
CrCl, ml/min	136 ± 7	157 ± 8	147 ± 4	183 ± 6
PiCl, ml/min	90 ± 14	115 ± 23	123 ± 7	149 ± 17

For significance see the text.

Table 4. Data concerning the patients on high UF HDF with two different PA areas

Parameters	PA 1.3 m^2	PA 1.6 m^2
Number	13	5
Sex	F = 8, M = 5	M = 5
Age, years	63 ± 12	45 ± 14
Postdialytic BW, kg	62 ± 7	68 ± 9
Dialytic age, months	35 ± 21	74 ± 48
Dialytic age on PA, months	21 ± 8	12 ± 3
Session length, min	215 ± 13	220 ± 10
Substituted volume, liters	22 ± 1	22 ± 3
Intradialytic BW loss, kg	2.2 ± 0.4	2.3 ± 0.3
Postdialytic Hb, g/dl	11.6 ± 1.1	12.8 ± 2.5
Predialytic BUN, mg/dl	67 ± 9	67 ± 11
Predialytic sCr, mg/dl	8.6 ± 0.8	11.1 ± 1.1

Table 5. Mean value ± SD of parameters examined in 9 patients transferred from hemodialysis with cuprophan (Cu-HD) to high UF HDF with PA (PA-HDF)

Parameters	Cu-HD	PA-HDF	p
Membrane area, m^2	1.6 in 6 patients 1.3 in 3 patients	1.6 in 4 patients 1.3 in 5 patients	
Session length, min	245 ± 15	220 ± 14	< 0.01
Postdialytic BW, kg	65 ± 8	65 ± 7	NS
Predialytic BUN, mg/dl	71 ± 8	63 ± 10	< 0.02
Predialytic sCr, mg/dl	11.3 ± 1.9	9.7 ± 1.2	< 0.005
Postdialytic RBCs, $1{,}000/mm^3$	3,700 ± 700	4,200 ± 780	< 0.001
Postdialytic Hb, g/dl	11.1 ± 2	12.9 ± 2	< 0.005

with other synthetic membranes ($p < 0.025$ for AN69, $p < 0.02$ for PAN, $p < 0.005$ for PMMA). No significant difference was found among the various membranes or between the two techniques for urea and creatinine clearance [7].

Comparison between 1.3 and 1.6 m² Polyflux in HDF Mode

The values of UF and small solute clearance obtained with the two Polyflux (PA) dialyzers are shown in table 2. Both the ultrafiltration rate and the clearances were significantly higher with the 1.6 m^2 Polyflux compared to the 1.3 m^2 Polyflux.

Comparison between HDF and HD Using Polyflux

The values of UF and small solute clearances by using the two Polyflux dialyzers in HD and HDF are shown in table 3. The clearances were significantly higher in HDF mode than in HD mode for both devices ($p < 0.01$ for UrCl, $p < 0.001$ for CrCl, $p < 0.05$ for PiCl).

Long-Term Evaluation of Patients on High UF HDF with Polyflux and Comparison with the Previous Treatment Mode

Table 4 shows clinical and dialysis data of 18 patients on long-term HDF with 1.3 m^2 and 1.6 m^2 Polyflux (PA). In addition, clinical and biochemical results following the transfer of patients from Cu-HD to Polyflux-HDF are shown in table 5.

Predialysis serum urea (BUN) and creatinine levels decreased significantly at the end of the observation period on Polyflux-HDF, despite a significant reduction of the session length, a constant dry body weight, and a comparable dialyzer surface area. A significant increase of red blood cell (RBC) count was also noted (the only patient on erythropoietin therapy was excluded from the RBC evaluation).

Discussion

All the studied high-flux dialyzers showed similar clearance of small solutes in spite of varying ultrafiltration rate. At low TMP the Polyflux gave significantly higher UF rates than the other dialyzers. In this comparative study, however, the surface area of the dialyzers with the polyamide and the AN69 membranes was smaller, 1.6 m^2, than those of the other dialyzers (1.8 and 2.0 m^2, respectively). This indicates a more efficient removal with these two former membrane types. The influence of the membrane surface area on the performance was well illustrated by the significant improvement in small molecule removal when going from the 1.3 to the 1.6 m^2 Polyflux dialyzer. Thus, a further increase of the surface area would be expected to give even higher clearances, provided the blood flow rate is sufficiently high.

When the same size Polyflux dialyzer was used in the same patient for hemodialysis and high ultrafiltration HDF, keeping a constant session length, blood flow and dialysate flow, it was found that HDF with high ultrafiltration produced a higher small molecule clearance than HD, confirming that diffusion and convection in combination give a better small molecule removal than each process used separately [2].

However, it is not yet established how much each single process contributes to the overall small solute clearance in HDF [5], or how far UF can be reduced without affecting such clearances. In fact, similar small molecule clearance was obtained in HDF in spite of different ultrafiltration rates (techniques A and B, table 1).

The long-term benefits seen with Polyflux used in high UF HDF were encouraging. The control of BUN and creatinine, together with the maintenance of dry body weight and the increase in hemoglobin, suggest that adequate dialysis was being prescribed despite of the significant reduction in treatment time. This could be due to the better removal of the middle molecules as well as the improved biocompatibility of the membrane leading to reduced protein catabolism [9–11].

Conclusions

High UF HDF with synthetic high-flux membranes guarantees an optimal removal of small as well as larger molecules and an improved biocompatibility, whilst avoiding the risk of backfiltration. Polyflux was shown to have a high efficiency comparable to other synthetic membranes. Polyflux dialyzers which are equipped with a polyamide membrane were shown to have a high efficiency regarding both ultrafiltration and solute clearance and gave excellent clinical results.

References

1 Ronco C: Backfiltration in clinical dialysis: Nature of the phenomenon, mechanisms and possible solutions. Int J Artif Organs 1990;13:11–21.
2 Gonella M, Pratesi G, Baldin C, Calabrese G, Vagelli G, Mazzotta A: Efficacy of hemodiafiltration vs hemofiltration and hemodialysis on small solutes' removal. Dial Transplant 1990;19:420–422.
3 Vaage R, Cavatorta F, Queirolo C, Rosselli P, Gentile A, Bertulla A: Hemodynamic changes during acetate dialysis, bicarbonate dialysis and hemofiltration. Blood Purif 1988;6:43–50.
4 Gonella M, Mazzotta A, Pratesi G, Calabrese C, Vagelli G: Emodiafiltrazione versus emodialisi nel paziente a rischio; in Minetti L, Orlandini GC, Petrella E (eds): L'emodiafiltrazione ed i trattamenti ad alti flussi. Milano, Wichtig Editore, 1989, pp 313–318.
5 Colton CK: Analysis of membrane processes for blood purification. Blood Purif 1987;5:202–251.
6 Chanard J, Toupance O, Lavaud S, Melin J: Long-term dialysis with highly permeable membranes: 15 years experience. J Nephrol 1989;3:213–219.
7 Gonella M, Pratesi G, Baldin C, Calabrese G, Vagelli G, Mazzotta A: The behavior of ultrafiltration rate and small molecules clearances in hemodiafiltration with four high-permeability membranes. Dial Transplant 1990;19:625–629.
8 Van Geelen JA, Flendrig JA, Carpay WA: Evaluation of in vivo solute clearances during hemodialysis and hemodiafiltration. Dial Transplant 1985;14:85–90.
9 Gonella M, Pratesi G, Calabrese G, Vagelli G, Mazzotta A: Improvement of anemia in patients on chronic dialysis treated by hemodiafiltration. Blood Purif 1989;7: 186–191.
10 Schaefer RM, Gilge U, Göhl H, Heidland A: Evaluation of a new polyamide membrane (Polyflux 130) in high-flux dialysis. Blood Purif 1990;8:23–31.
11 Dinarello CA: Interleukin 1. Its multiple biological effects and its association with hemodialysis. Blood Purif 1988;6:164–172.

Dr. M. Gonella, Ospedale S. Spirito/Dialisi, Viale Giolitti 2,
I-15033 Casale Monferrato, AL (Italy)

Shaldon S, Koch KM (eds): Polyamide – The Evolution of a Synthetic Membrane for Renal Therapy. Contrib Nephrol. Basel, Karger, 1992, vol 96, pp 106–110

High-Flux Hemodialysis with a Polyamide Dialyzer

Claude Jacobs, Thierry Petitclerc[1]

Département de Néphrologie, Groupe Hospitalier Pitié-Salpêtrière, Paris, France

The performance of the Gambro Polyflux 130 dialyzer (1.3 m^2) equipped with a hydrophilic polyamide membrane has been evaluated in a series of 80 hemodialysis sessions. Removal and clearance of low-molecular-weight substances have been measured, and in addition clinical tolerance was evaluated. The results have proven very satisfactory from the clinical and biochemical standpoint. These results illustrate the performance of the device when used in the conventional hemodialysis mode with mainly diffusive transport.

Patients and Methods

The dialysis sessions were performed in 30 stable long-term dialysis patients who all had adequate vascular accesses punctured with two needles. Blood flow was routinely set at 200 ml/min as displayed on the BSM22 blood pump on the Monitral S (Hospal) dialysis delivery system that was used throughout the study. The accuracy of the blood flow indicated by the blood pump was validated by blood flow measured with an air-bubble transit time. The dialysate composition was (mmol/l): sodium 140, potassium 2, bicarbonate 31. The dialysate flow rate was 500 ml/min. The duration of the dialysis session was 4 h in 73 sessions and 3 h 30 min in 7 sessions. Anticoagulation was performed using conventional intermittent heparinization of the blood circuit (initial bolus 1,800 IU, mean maintenance dose 3,250 IU) in 15 sessions. In 65 sessions, a single injection of low-molecular-weight heparin (average dose 50 IU/kg BW) was given at the start of dialysis. The ultrafiltration (UF) rate was set for each dialysis session according to the weight

[1] The authors are indebted to Mrs. F. Raymond for skillful technical assistance and to Mrs. C. Debrun for preparing the manuscript.

loss required to reach the patient's dry weight at completion of dialysis. The mean UF rate (Q_f) was 13.7 ml/min. Arterial blood samples for determination of plasma urea and creatinine concentrations with standard laboratory techniques (Technicon®) were drawn pre-and post-dialysis at each of the hemodialysis (HD) sessions. Arterial and venous blood samples were drawn 120 min after the start of dialysis in 52 sessions for determination of the dialyzer clearances.

Calculations

The Polyflux dialyzer clearance for urea and creatinine was calculated on the 'blood-side' of the dialyzer using the formula:

$$\text{Clearance} = (Q_b \times E) + Q_f(1 - E),$$

Q_b = blood flow (ml/min); Q_f = filtration rate (ml/min):

$$E = \frac{C_{\text{arterial}} - C_{\text{venous}}}{C_{\text{arterial}}},$$

C = concentrations of urea or creatinine at the arterial and venous ends of the dialyzer. The post/pre-dialysis reduction of arterial plasma concentration of urea and creatinine (PR) was calculated for the 80 sessions using the formula:

$$PR = \frac{C_o - C_t}{C_o} \times 100,$$

C_o = concentration at start of dialysis; C_t = concentration at end of dialysis. Kt/V for urea was calculated with the formula:

$$\frac{Kt}{V} = \ln \frac{C_o}{C_t}.$$

Statistical analyses were done using the Student's t test with a level of significance set at 5%. All values are given as mean ± SD.

Results

The clearance of the Polyflux dialyzer for urea and creatinine measured during 52 dialysis sessions, 2 h after start of dialysis with Q_b 200 ml/min, Q_d 500ml/min (dialysate flow) and mean Q_f 13.7 ml/min were C_{urea} 143 ± 16 ml/min and $C_{creatinine}$ 116 ± 11ml/min. If the UF rate was extrapolated to zero, the clearance values for urea and creatinine changed to 136 and 109 ml/min. When Q_b was increased to 240 ml/min, C_{urea} rose to 170 ml/min and $C_{creatinine}$ rose to 156 ml/min. The percentage reduction in plasma concentration (post/pre-dialysis) was 60 ± 10% for urea and 55 ± 6% for creatinine.

The duration of dialysis, 3 h 30 min compared to 4 h, did not change the percentage reduction of these small molecules between the different heparinisation protocols. With intermittent heparinization (n = 15), PR_{urea} was 58.0 ± 8.0% and with low-molecular-weight heparin (n = 65), PR_{urea} was 60.3 ± 8.5% (p = NS). Similarly, $PR_{creatinine}$ was 53.0 ± 3.0% for intermittent heparin and 56.4 ± 7.0% with low-molecular-weight heparin (p = NS). The mean value of Kt/V urea for the 73 sessions of 4 h duration was 1.10 ± 0.1.

Clinical Tolerance

No adverse side effects attributable to the dialyzer were observed durig the 80 sessions reviewed in this study. The patients' average weight loss was 2.6 ± 1.2 kg during the 4-hour dialysis sessions and 2.1 ± 1.2 kg during the shorter sessions. Five episodes of symptomatic hypotension occurred in 5 different patients. They were easily corrected by intravenous infusion of 200 ml of isotonic saline (2 patients) or 200–250 ml of a macromolecular solution (Plasmagel) (3 patients). There was one blood leak that occurred at 30 min after the start of dialysis. The cause of this incident remained undetermined. Blood restitution to the patients at completion of dialysis was deemed satisfactory on visual criteria with either anticoagulation technique.

Comments

Several studies have emphasized the excellent results obtained in vitro as well as in vivo with hydrophilic polyamide dialyzers, particularly concerning the removal of β_2-microglobulin [1–3]. When used for hemofiltration, the hydrophilicity of the membrane explains the optimization of the sieving coefficient for β_2-microglobulin. This reflects not only a lesser degree of adsorption on the membrane surface but also the potential for a higher total convective mass removal [1–3]. Up to 69% reduction of plasma β_2-microglobulin concentration has been reported in 4 h using the polyamide membrane in the dialysis modality [3]. The hydrophilic polyamide membrane also yielded very satisfactory results on various biocompatibility parameters, such as the degree of leukopenia during the early phase of a dialysis session, complement and leukocyte activation and protein loss [2]. In view of the excellent properties of the polyamide membrane regarding biocompatibility and removal of β_2-microglobulin, it was of great interest to study the removal of the small solutes, traditionally used as surrogate markers of uremic toxicity. The purpose of the present study was therefore to assess the efficacy of the Polyflux 130 dialyzer on the clearance of urea and creatinine as well as its clinical tolerance in a population of maintenance hemodialysis patients. Under the conditions of utilization (blood flow 200 ml/min), it is of interest that the mode of heparinization did not affect the percent reduction of urea or creatinine. It should be noted that there was a discrepancy between our in vivo clearance values for urea and creatinine (C_{urea} 143 ± 16 ml/min and $C_{creatinine}$ 116 ± 11ml/min) and those obtained in vitro and published by the manufacturer (C_{urea} 181 ml/min and $C_{creatinine}$ 170 ml/min). This can be accounted for by the latter determinations being made with an UF rate of 60 ml/min, i.e. clearly in a hemodiafiltration mode. It has been shown that with increasing UF the clearance of low-molecular-weight solutes is increased by about 50% of the value of the UF [4]. At zero UF the manufacturer's recorded value of urea clearance in vitro would thus be 151 ml/min and our own in vivo average value was 136 ml/min. Similarly, the in vitro creatinine clearance as measured by the manufacturer would be 140 ml/min and in our in vivo study it was 109 ml/min.

In view of this difference, we checked the possibility of an error in our blood flow measurement. We found that the bubble transit time method yielded a flow 20% higher than the Q_b used in the clearance calculations, which was taken from the blood pump reading. Our clearance values were thus recalculated using the higher blood flow, i.e. 240 ml/min. With this correction being taken into account and with extrapolation to zero UF rate, C_{urea} became 170 ml/min and $C_{creatinine}$ became 156 ml/min. These clearances were within the range of the in vitro values given by the manufacturer for blood flows of 200 and 300 ml/min.

These results clearly show that the performance of the Polyflux 130 dialyzer used in the conventional hemodialysis mode is within the same range of clearances of low-molecular-weight solutes as reported in the literature for other high-flux and standard dialyzers of comparable surface area [5].

The clinical tolerance and patients' acceptance of this dialyzer were rated as excellent in this short-term study. The results of the in vivo investigations reported here, as well as elsewhere, highlight the very attractive features of this filter for treating chronically uremic patients in the conventional hemodialysis mode.

However, as is the case with all materials used for renal replacement therapy, only long-term, extensive clinical experience will permit more definitive and substantiated statements on the advantages and cost-effectiveness of the hydrophilic polyamide dialyzer.

References

1 Floege J, Granolleras C, Smeby L, Göhl M, Koch KM, Shaldon S: Hydrophilic high-flux polyamide membranes for beta-2-microglobulin removal. Trans Am Soc Artif Intern Organs 1987;33:309–311.
2 Shaefer RM, Gilge U, Göhl H, Heidland A: Evaluation of a new polyamide membrane (Polyflux 130) in high-flux dialysis. Blood Purif 1990;8:23–31.
3 Floege J, Granolleras C, Deschodt G, Heck M, Baudin G, Branger B, Tournier O, Reinhard B, Eisenbach GM, Smeby L, Koch KM, Shaldon S: High-flux synthetic versus cellulosic membranes for beta-2-microglobulin removal during hemodialysis, hemodiafiltration and hemofiltration Nephrol Dial Transplant 1989;4:653–657.
4 Jaffrin MY, Ding LH, Laurent JM: Contribution of convection to hemodiafiltration clearance. Blood Purif 1989;7:266–267.
5 Hoenich NA, Woffindin C, Ward MK: Dialyzers; in Maher JF (ed): Replacement of Renal Function by Dialysis. Dordrecht, Kluwer, 1989, pp 144–180.

Prof. C. Jacobs, Service de Néphrologie, Hôpital de la Pitié,
83, Boulevard de la Pitié, F-75651 Paris Cedex 13 (France)

Shaldon S, Koch KM (eds): Polyamide – The Evolution of a Synthetic Membrane for Renal Therapy. Contrib Nephrol. Basel, Karger, 1992, vol 96, pp 111–123

Continuous Renal Replacement Therapies of Acute Renal Failure and the Polyamide Membrane

Claudio Ronco

Department of Nephrology, St. Bortolo Hospital, Vicenza, Italy

Continuous arteriovenous hemofiltration (CAVH), invented by Kramer in 1977 [1] and later developed by others [2], has been widely used for the treatment of acute renal failure in critically ill patients [2–4]. CAVH is an extracorporeal treatment which removes water, electrolytes and metabolic waste products from the body by ultrafiltration. The partial or total replacement of the ultrafiltered fluid with different substitution solutions lowers the blood concentration of several toxins and gives a desired body weight and electrolyte balance resulting in a general equilibration of the patient [5]. In the pioneering era, the technique frequently presented problems of low efficiency and clotting of the circuit resulting in a high percentage of treatment failures. This was mainly due to limited experience and to the nonspecific devices available. In recent years, new products specifically designed for continuous therapy have become available and permitted a definite improvement in the quality of the treatment [6, 7]. Special circuits and filters as well as suitable vascular accesses and catheters have been created to make the treatment simpler and safer [8]. In this field, the polyamide membrane has been utilized quite extensively and two special hemofilters, FH66 and FH22 (Gambro AB), have been developed for continuous therapies [7]. These two filters are now in regular use. Our technical and clinical experience of continuous renal replacement therapy with them will be presented.

The Extracorporeal System

CAVH as originally described consisted of a hemofilter connected to the femoral artery and a vein by short blood tubing sets and special percutaneous catheters [1]. Several modifications of the original technique have been proposed and a large variety of derived techniques is available today, including continuous arteriovenous hemodialysis (CAVHD), continuous veno-venous hemofiltration (CVVHF) and continuous veno-venous hemodialysis (CVVHD) (fig. 1) [9–11]. In standard CAVH, the access to the circulation must provide an adequate arteriovenous pressure gradient in order for the blood to flow through the circuit [8]. Since no pumps are used, the resistance of the access should be reduced to the minimum in order to achieve the maximal blood flow. For this reason shortened catheters or standard catheters with a large inner diameter must be utilized for the percutaneous cannulation of the femoral vessels. The blood lines should also be short for the same reason. When a Scribner shunt is employed, large bore tips are recommended for the cannulation of the vessels of the arm. More and more a veno-venous circulation is preferred and double lumen venous catheters are utilized with a blood pump [6–8]. Both systems (A-V or V-V) have advantages and disadvantages, and the right way to look at the problem is to consider both possibilities as a possible choice. Since no special features are required for the hemofilters used in a pump-driven circulation, the spontaneous arteriovenous mode will be discussed. When no pumps are used, the driving force for filtration, i.e. the transmembrane pressure, is remarkably low and it could even be insufficient to achieve satisfactory ultrafiltration rates [12]. In this case, the ultrafiltrate line must be positioned to create a suction pressure favoring filtration. Therefore, the height of the column of ultrafiltrate is critical to ensure an adequate range of transmembrane pressure.

Hemofilters

Filters used for continuous therapies utilize membranes of polyamide, polysulfone and polyacrylonitrile. Synthetic membranes have in fact been considered as the material of choice for these treatments. The evolution of the asymmetric polyamide membrane has permitted the development of two specific hemofilters: the FH66, 0.6 m^2, for adults, and the FH22, 0.22 m^2, for children or small adults. Special care has been taken with the

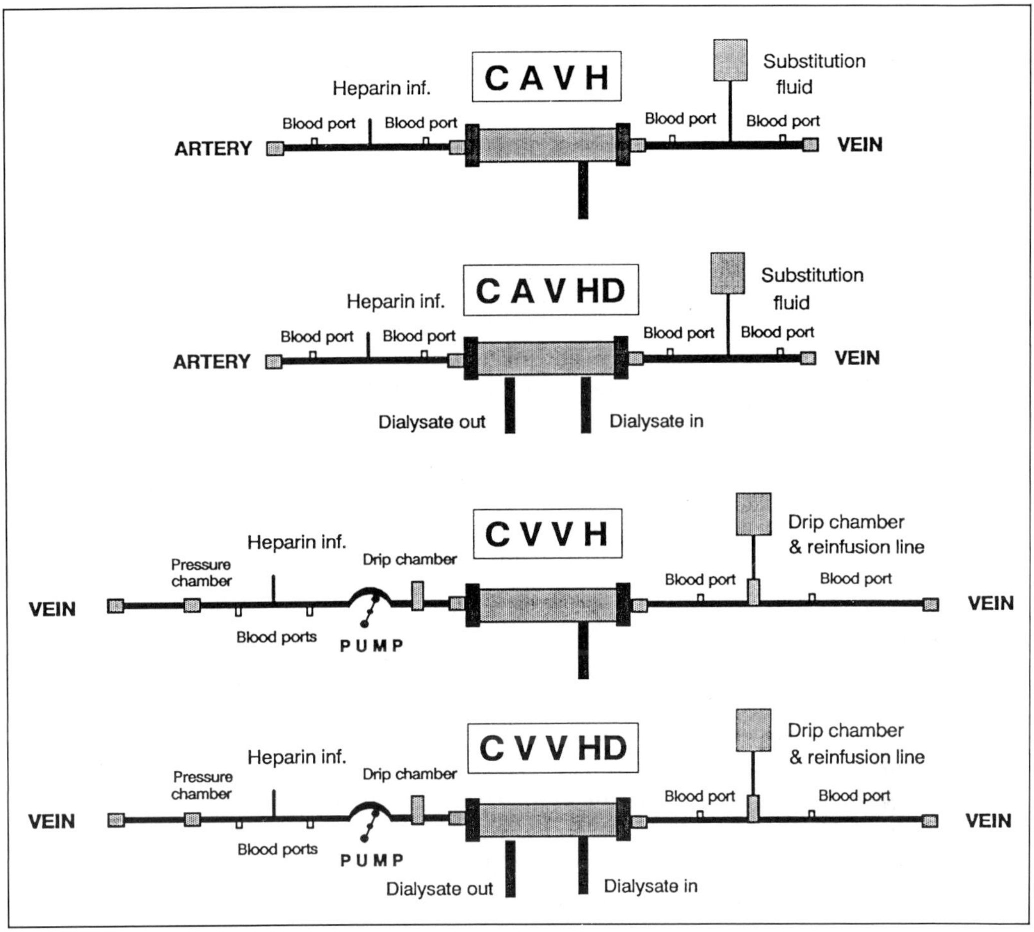

Fig. 1. Different configurations of the extracorporeal circuit for differing continuous renal replacement therapies. The blood flow rate in CAVH and CAVHD is a function of the A-V pressure gradient and of the resistance in the circuit.

design of these devices in order to achieve optimal performance not only in pumped circulation but also in an arteriovenous spontaneous driven mode. In figure 2, the hydraulic properties of the two devices derived from in vitro data with pumped blood are reported in detail. The blood flow rates achieved at different perfusion pressures generated by gravity are remarkably high and linearly correlated (fig. 2a). This correlation describes the intrinsic resistance of the devices which appears to be rather low. The difference between the two filters depends on the different blood path geometry, dimension and surface area as shown in table 1. With a progressive increase of the blood flow obtained with a blood pump, the pressure drop inside the filters tends to increase, but the recorded values demonstrate a remarkably low resistance in the blood path (fig. 2b). Again the smaller cross-sectional area with a similar length accounts for the somewhat larger pressure drop in FH22. The spontaneous filtration rates achieved in response to blood flow without additional pressures in the range of 20–30 ml/min for the FH66 and 3–7 ml/min for the FH22 (fig. 2c). When TMP is mechanically increased, there is a linear correlation between ultrafiltration rate and pressure up to 200 mm Hg. The filtration reaches a plateau value in the range of 45 ml/min for FH66 and 10 ml/min for FH22, respectively (fig. 2d). These hydraulic features are of great advantage in a spontaneous arteriovenous driven circulation [12]. The low resistance of the devices and the high permeability of the membrane yield an optimal performance of the filters even at blood flow rates below 100 ml/min. This gives a specific advantage in an arteriovenous circulation, where higher blood flows at a given arteriovenous pressure gradient permit reduction of the heparin dose and prolongation of the filter life span. With the FH66 device the phenomenon of filtration pressure equilibrium can be avoided in a wide range of clinical conditions. This plays an important role in the maintenance of the filter function and treatment success.

It appears evident that the two devices should be used for different purposes. The FH66 must be utilized when high ultrafiltration rates are required and large amounts of fluid should be exchanged in the patient. The FH22 (where the priming volume and filtration rate are lower) is particularly indicated for small size adults, children or severely overhydrated patients in whom slow continuous ultrafiltration without a replacement solution is required. The use of polyamide and other synthetic membranes achieves satisfactory removal of larger molecules due to membrane-sieving properties and convective transport [13]. When an increased clearance of small molecules like urea is needed, diffusion can be added to the standard convective treatment by circulating dialysate in the ultrafiltrate compartment (fig. 1) [14]. This procedure is called continuous hemodialysis, or continuous hemodiafiltration if a high-flux membrane and a reinfusion solution are used [15].

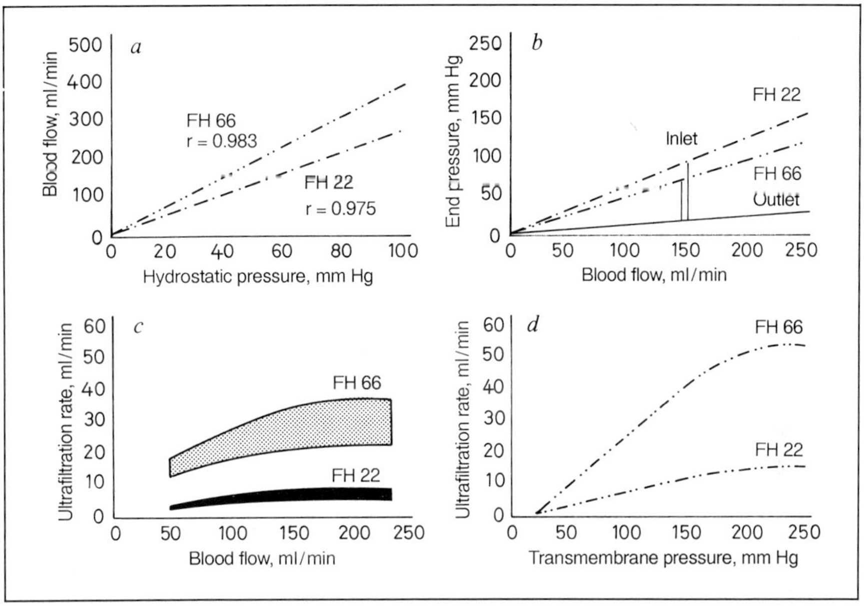

Fig. 2. Hydraulic characteristics of the FH66 and FH22 devices in vitro derived from pumped blood. *a* Blood flow through the device at various perfusion pressures. *b* End-to-end pressure drop at various blood flows (hematocrit = 30 vol% and total protein = 6 g/dl). *c* Spontaneous filtration at various blood flows without any additional transmembrane pressure. *d* Ultrafiltration rate vs. transmembrane pressure at a constant blood flow of 150 ml/min.

Table 1. Characteristics of polyamide filters for CAVH

	FH66	FH22
Surface area, m^2	0.6	0.16
Length of the unit, mm	180	150
Effective length of the fibers, mm	140	115
Diameter of the unit, mm	38	29
Priming volume, cm^3	43	11
Membrane thickness, μm	50	60
Hollow fiber diameter, μm	215	215
Number of fibers	6,200	2,100
Residual blood volume, ml	<1	<1
Maximal TMP, mm Hg	600	600
In vitro Q_f at TMP = 100, ml/min	20–26	4–5

Clinical Experience

CAVH was initially proposed as an alternative treatment for acute renal failure when hemodialysis or peritoneal dialysis were contraindicated. Today, CAVH and derived therapies are considered as a first-choice treatment in patients with complicated forms of acute renal failure, refractory overhydration, severe cardiovascular instability, fluid and electrolyte derangements and multi-organ failure. The reason is the excellent clinical tolerance, the easy monitoring and the low rate of complications [16, 17]. Since 1987, we have treated 26 patients with continuous therapies using the polyamide membrane in either the FH66 or FH22 configuration. The population treated and the relevant parameters are reported in table 2. It can be noted that none of the patients presented with uncomplicated acute renal failure and all of them had important medical or surgical complications. The rate of survival is similar to other reports and it can be considered satisfactory when related to the severity of organ failure in the individual case. It is evident that while in the multiple organ failure syndrome there were no survivors, the rate of success for other forms of complicated acute renal failure (ARF) is more encouraging (e.g. polytraumatic ARF, 3 out of 5 survivors). The majority of the patients were treated with CAVH and in 2 cases only was CAVHD employed. Percutaneous femoral cannulation was the access of choice. In 2 patients a double lumen catheter implanted in the femoral vein was used with a blood pump. In 2 other patients a forearm Scribner shunt was inserted. In the neonates the femoral vessels were cannulated after a surgical cut down procedure. In such small size patients the FH22 device was used and satisfactory blood flows and ultrafiltration rates could be achieved (table 3). With the FH66 device, an average ultrafiltrate volume of 15.5 liters/24 h permitted adequate blood urea nitrogen control in the majority of the adult patients. When a severe hypercatabolic state required an augmented clearance of BUN (2 patients), diffusion was added to the convective transport and the treatment was carried out in the CAVHD mode. In all patients, an adequate control of body weight could be achieved as well as a satisfactory electrolyte and acid-base balance. This was made possible by accurate fluid balance monitoring and an appropriate choice of substitution fluid. The operational characteristics obtained during the treatment of the 26 patients are shown in table 3. In figure 3, some data recorded from 9 CAVH patients treated with the FH66 device are reported to show in detail the average performance of the filter over 24 h of use. It can be noted that there was a pro-

Table 2. Diagnosis and outcome of the treated population

Diagnosis	Treated	Survived
Multiple organ failure	2	0
ARF polytrauma	5	3
ARF postsurgery	11	6
Hemorrhagic shock	2	2
Congestive heart failure	2	2
ARF in newborns	4	2
Total	26	15

ARF = Acute renal failure.

Table 3. Operational characteristics of polyamide filters in CAVH (mean ± SD)

	FH66	FH22
Patients treated	22	4
Sex (M/F)	13/9	2/2
Age	47 ± 13 (years)	5 ± 3 (days)
Mean arterial pressure, mm Hg	87 ± 23	45 ± 9
Blood flow, ml/min	84 ± 37	28 ± 6
Ultrafiltration rate, ml/min	11 ± 4	2 ± 1
Filter life span, h	46 ± 14	41 ± 21
Heparin administration, IU/h	600 ± 250	25 ± 10
Total ultrafiltrate, liters/24 h	15.5 ± 2.1	1.3 ± 0.3
Total reinfusion, liters/24 h	12.1 ± 3.2	1.2 ± 0.4
Treatment duration, days	9 ± 5	5 ± 3

gressive decrease of the ultrafiltration rate, accompanied by a parallel reduction of the blood flow rate. The filtration fraction remained unchanged, varying between 16 and 18%. This behavior suggests clotting of some fibers during the treatment with a slight reduction of the available surface area. However, the sieving coefficients of both urea and inulin remained stable over the 24-hour study period, suggesting that the permeability of the membrane remained fairly constant. This phenomenon might be explained by the membrane's hydrophilic nature leading to lower interaction with plasma proteins. No special complications occurred in

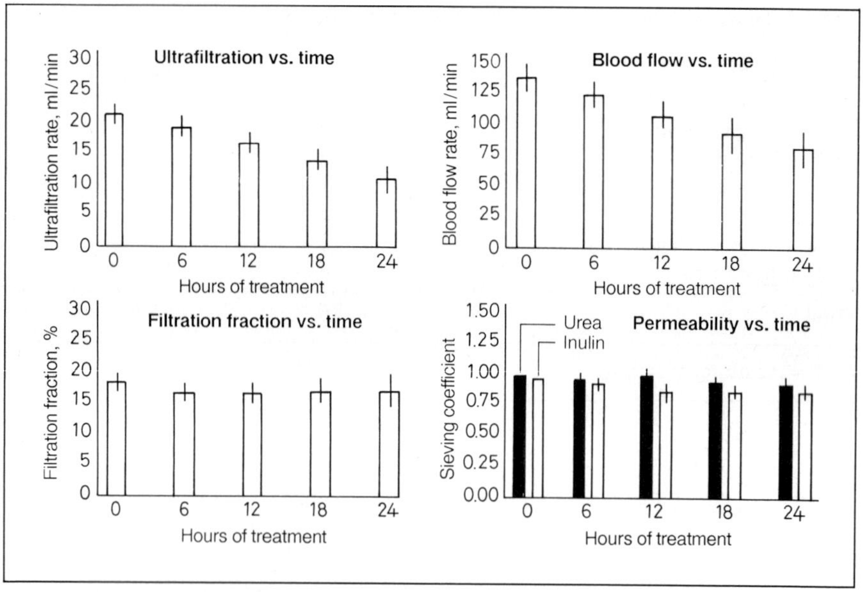

Fig. 3. Operational characteristics of the FH66 device recorded in 9 patients of a period of 24 h (mean ± SD).

relation to the continuous therapy or with the extracorporeal circulation. A remarkable stability of the patients' blood pressure could be observed in all cases. Platelet counts and clotting times were maintained within acceptable ranges through accurate and continuous monitoring. In 2 cases, hypothermia was observed and the use of warmed substitution fluid corrected this problem. In 1 patient, the arterial catheter had to be replaced after 6 days of treatment because of severe bleeding around the catheter. In patients who died, the causes of death were not related to the treatment procedure or to their renal failure.

Clinical Aspects of Continuous Therapies

The clinical advantages of continuous therapies in the treatment of complicated ARF can be summarized as follows: Easy institution and monitoring, no machine requirement, optimal clinical tolerance, and satisfactory treatment efficiency. The clinical conditions in which continuous

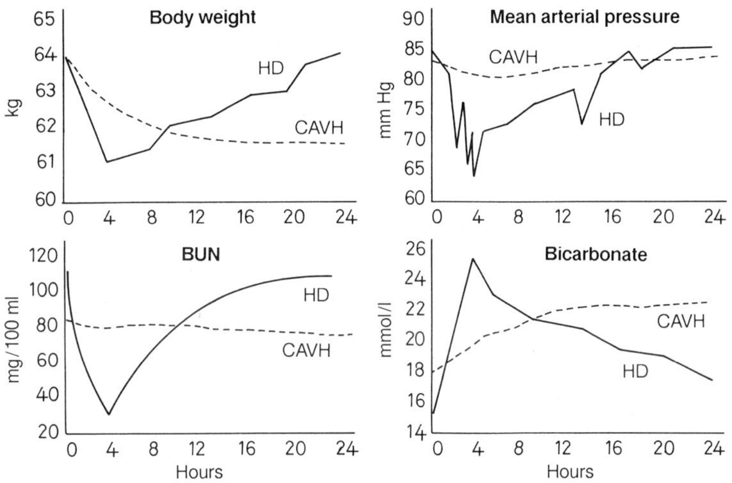

Fig. 4. Comparison of the clinical response and biochemical profile of a typical patient during a 24-hour period, when he was treated either with intermittent hemodialysis (HD 4 h) or with CAVH continuously. The CAVH period demonstrates a constancy of blood chemistry and body weight as well as a marked stability of the mean arterial blood pressure.

therapies are particularly indicated are: fluid overload, accumulation of uremic toxins, electrolyte and acid-base derangements in patients with severe cardiovascular instability, multiple organ failure and surgical complications. In these conditions, continuous therapies can display a unique behavior not achievable with other techniques [18]. Comparison of the clinical response and biochemical profile of a typical patient with ARF treated by daily intermittent hemodialysis for 3 days and then by CAVH for 9 days is shown in figure 4. The CAVH period demonstrates a constancy of blood chemistry and body weight as well as a marked stability of the mean arterial blood pressure.

CAVH allows a continuous progressive fluid removal from the patient which is generally well tolerated. Several factors have been proposed to explain the vascular stability during CAVH. These include the slow continuous ultrafiltration, the plasma refilling due to iso-osmotic production of ultrafiltrate and the stability of the renin-angiotensin system. A great advantage of CAVH is linked to the possibility of dissociating the removal of sodium and water. It is in fact possible to remove sodium and water independently by changing the composition of the replacement solution. This effect may influence the behavior of the peripheral resistance during CAVH. The hemodynamic response to fluid withdrawal by CAVH may be different from that obtained by other extracorporeal therapies as the stability of the arterial blood pressure is well recognised. Vascular stability makes CAVH the first choice treatment in patients with ARF associated with myocardial insufficiency and congestive heart failure and may result in an improvement of cardiac function during therapy. CAVH can be successfully used as an alternative treatment in patients with ARF who cannot be treated with other therapies because of medical or technical problems, such as patients with multi-organ failure or after open heart surgery. Should a severely catabolic state be present, several supporting procedures can be applied to improve the efficiency of the system in removing low molecular waste products: (a) The optimization of the extracorporeal circuit ensures the maximal performance of the treatment and its continuous action. (b) An adequate hyperalimentation can help reduce the protein catabolic rate and thus have an effect on BUN generation. (c) When high blood viscosity or low blood flows are present, the use of predilution rather than postdilution can be recommended. The consequent increase in ultrafiltration and BUN clearance can be useful to control the BUN concentration. Furthermore, the heparin requirement can be reduced. (d) Although the new generation of hemofilters allows for ultrafiltration rates ranging between 10 and 28 ml/min, sometimes the patient's hypotension does not allow for blood flows adequate to achieve the best performance of the system, and the BUN level cannot be maintained under control. In these patients the efficiency of the system can be further increased by adding diffusion to the convective transport. Other strategies such as vacuum suction in the UF compartment or the use of a blood pump are still controversial. They may not represent real advantages as they make the treatment more difficult and complicated.

CAVH may be used to correct water and electrolyte imbalances by varying the composition of the substitution fluid. Hyponatremia and hypernatremia can be corrected not only by achieving a normal plasma sodium concentration, but also by restoring the normal body sodium content. This feature renders CAVH particularly indicated in patients with overhydration, hyponatremia and high total body sodium. Hyperkalemia can also be corrected with CAVH; the efficiency in removing potassium is directly related to the amount of fluid removed during the treatment and its replacement with potassium-free solutions. Calculation of ultrafiltration rate, amount and composition of the replacement solution and blood flow may permit a theoretical prediction of the time necessary for the correction of the patient's electrolyte imbalance. Bicarbonate loss during CAVH can easily be measured directly in the ultrafiltrate or predicted using the formula:

$$HCO_3(f) = UF \times HCO_3(s) \times 1.124,$$

where $HCO_3(f)$ = amount of bicarbonate in the ultrafiltrate; $HCO_3(s)$ = bicarbonate concentration in the serum; UF = total volume of ultrafiltrate, and 1.124 = average sieving coefficient for bicarbonate.

When CAVH is applied without substitution fluid in order to reduce the patient's fluid overload, bicarbonate losses are compensated by the reduction of the body volume distribution for the buffer and the plasma concentration does not change significantly. On the contrary, when replacement solutions are infused to maintain the body fluid balance, the same amount of bicarbonate lost in the ultrafiltrate must be administered to achieve stable plasma levels of the buffer. Finally, when CAVH is utilized to correct metabolic acidosis, the amount of bicarbonate in the replacement solution must exceed the amount lost in the ultrafiltrate, providing a positive balance of the buffer. In conclusion, when a correction of water, electrolyte or acid-base derangement is required, the rate of ultrafiltration, the rate of reinfusion, the initial plasma concentration, the amount lost in the ultrafiltrate, the content of the replacement solution and the rate of metabolic production must all be considered for the different solutes in order to achieve the desired final balance and plasma concentration.

CAVH in Newborns

The recent introduction of the FH22 device has allowed small babies and newborns to be treated with CAVH. The treatment is generally well tolerated and an ultrafiltration of 1–3 ml/min is enough to ensure the adequate efficiency of the therapy in babies suffering from acute renal failure. Special blood lines and catheters are now available and the treatment is becoming more and more popular due to the excellent results observed [19].

Conclusions

The use of the above-mentioned procedures and devices using synthetic membranes renders CAVH a reliable and efficient technique for the treatment of patients with acute renal failure. Some specific advantages such as the simplicity, the easy monitoring and easy institution make this technique a first choice treatment in several clinical conditions. For patients with severe cardiovascular instability, multi-organ failure or polytraumatism, CAVH is today more than a simple alternative treatment to hemodialysis. The institution of the above-described procedures with new devices overcomes the classic disadvantages of CAVH such as the low clearance efficiency or the frequent clotting of the filters. Critically ill patients should be treated increasingly by this technique in the future. In this regard, the polyamide membrane has found an excellent application and represents an important step forward in progress for all forms of continuous renal replacement therapy.

References

1 Kramer P, Wigger W, Rieger J, Matthaei D, Scheler F: Arteriovenous hemofiltration: a new and simple method for treatment of overhydrated patients resistant to diuretics. Klin Wochenschr 1977;55:1121–1122.

2 Lauer A, Saccggi A, Ronco C, Belledonne M, Glabman S, Bosch JP: Continuous arteriovenous hemofiltration in the critically ill patient. Ann Intern Med 1983;99: 455–460.

3 Ronco C, Brendolan A, Bragantini L, Chiaramonte S, Feriani M, Fabris A, La Greca G: Continuous arterio-venous hemofiltration. Contr Nephrol. Basel, Karger, 1985, vol 48, pp 70–78.

4 Bartlett R, Bosch JP, Paganini EA, Geronemus R, Ronco C: Continuous arteriovenous hemofiltration. Trans Am Soc Artif Intern Organs 1987;38:345–349.

5 Paganini EP: Acute Continuous Renal Replacement Therapy. Boston, Martinus Nijhoff, 1986.

6 Golper TA, Ronco C, Kaplan AA: Continuous arterio-venous hemofiltration: Improvements, modifications and future directions. Semin Dialy 1988;1:50–54.

7 Ronco C, Brendolan A, Bragantini L, Fabris A, Feriani M, Charamonte S, Milan M, Dell'Aquila R, La Greca G: Technical and clinical evaluation of a new polyamide hollow fiber hemofilter for CAVH. Int J Artif Organs 1988;11:33–38.

8 Olbricht CJ, Schurek HJ, Tytul S, Muller C, Stolte H: Comparison between Scribner shunt and femoral catheters as vascular access for continuous arteriovenous hemofiltration; in Kramer P (ed): Arteriovenous Hemofiltration. Berlin, Springer, 1985, pp 57–66.

9 Kaplan A, Longnecker RE, Folkert VW: Continuous arterio-venous hemofiltration. Ann Intern Med 1984;100:358–364.

10 Golper TA, Kaplan AA, Narasimhan N, Leone M: Transmembrane pressures generated by filtrate line suction maneuvers and predilution fluid replacement during in vitro continuous arteriovenous hemofiltration. Int J Artif Organs 1987;10:41–46.

11 Ronco C, Brendolan A, Bragantini L, Chiaramonte S, Dell'aquila R, Fabris A, Feriani M, Milan M, La Greca G: Arteriovenous hemodiafiltration associated with continuous arteriovenous hemofiltration. A combined therapy in the hypercatabolic patient; in La Greca G, Fabris A, Ronco C (eds): Proceedings of the International Symposium on Continuous Arteriovenous Hemofiltration, Vicenza 1986. Milano, Wichtig, 1986, pp 171–183.

12 Pallone TL, Peterson J: Continuous arteriovenous hemofiltration, an in vivo simulation. Trans Am Soc Artif Intern Organs 1987;33:304–308.

13 Lysaght MJ: Hemodialysis membranes in transition. Contr Nephrol. Basel, Karger, 1988, vol 61, pp 1–17.

14 Geronemus R, Schnaider N: Continuous arteriovenous hemodialysis; in Paganini E, Geronemus R (eds): Proceedings of the Third International Symposium on Acute Continuous Renal Replacement Therapy, Ft. Lauderdale, 1987, pp 64–77.

15 Ronco C: Arterio-venous hemofiltration (AVHDF): A possible way to increase urea removal during CAVH. Int J Artif Organs 1985;8:61–62.

16 Ronco C, Brendolan A, Bragantini L, Chiaramonte S, Feriani M, Fabris A, La Greca G: Continuous arteriovenous hemofiltration with AN69S membrane; procedures and experience. Kidney Int 1988;24:S150–S153.

17 Davenport A, Roberts NB: Amino acid losses during continuous high-flux hemofiltration in the critically ill patient. Crit Care Med 1988;17:1010–1014.

18 Maher ER, Hart L, Levy D, Scoble JE, Baillod RA, Sweny P, Varghese Z, Moorhead JF: Comparison of continuous arteriovenous hemofiltration and hemodialysis in acute renal failure (letter). Lancet 1988;i:129.

19 Zobel G, Trop M, Ring E, Grubbauer HM: Suction-supported continuous arteriovenous hemofiltration in children. Blood Purif 1988;6:37–42.

Dr. C. Ronco, Ospedale S Bortolo/Nefrologia, Via S Bortolo 8,
I-36100 Vicenza (Italy)

Shaldon S, Koch KM (eds): Polyamide – The Evolution of a Synthetic Membrane for Renal Therapy. Contrib Nephrol. Basel, Karger, 1992, vol 96, pp 124–137

Dialysis-Related Amyloidosis and High-Flux Membranes

Jürgen Floege[a], *Leif Smeby*[b]

[a] Division of Nephrology, Medizinische Hochschule Hannover, FRG;
[b] Gambro AB, Lund, Sweden

Dialysis-related amyloidosis has recently been recognized as one of the major complications of long-term renal replacement therapy [1–3]. Using immunohistochemical and biochemical techniques it has been shown that the amyloid fibrils of dialysis-related amyloidosis consist mainly of β_2-microglobulin [4]. β_2-Microglobulin, an 11.8-kD protein, is the light chain of the HLA-class I complex [1, 2]. Following its dissociation from the HLA-complex, β_2-microglobulin is normally found in low concentrations (1–3 mg/l) in the plasma, which markedly increase in end-stage renal failure due to the lack of renal excretion of this low-molecular-weight protein [5]. This appears to be the basic precondition for the deposition of β_2-microglobulin in amyloid fibrils (see below). The prevalence of dialysis-related amyloidosis in patients treated for more than 15 years is in the range of 50–100% [1–3]. In contrast to AL or AA amyloidosis, the clinical manifestations of dialysis-related amyloidosis are largely confined to musculoskeletal tissues, and clinically apparent organ manifestations are exceptional [1, 2]. The syndrome is not only observed in patients on long-term hemodialysis, but has also been described in rare patients treated exclusively with long-term hemofiltration or peritoneal dialysis [6, 7]. As in other types of amyloidosis, the knowledge of the pathogenesis of dialysis-related amyloidosis is incomplete. Consequently, effective therapeutic or preventive strategies remain to be developed. However, in contrast to other types of amyloidosis two features of dialysis-related amyloidosis may facilitate the design of preventive and possibly even therapeutic measures.

Firstly, the disease appears to occur nearly invariably in end-stage renal disease patients, i.e. the patient at risk is known long before the onset of disease, and, secondly, dialysis-related amyloidosis appears to be largely derived from circulating β_2-microglobulin, i.e. the toxin is known long before the onset of disease. Since the amyloidosis is limited to end-stage renal disease patients, the ideal preventive strategy is the early restitution of normal renal function by a successful kidney transplantation. Support for this notion is derived from the observation that even in established cases of dialysis-related amyloidosis a successful kidney transplant could arrest the natural progression of the disease, although it was not therapeutic since it did not induce regression of the amyloid deposits [8]. In contrast to renal transplantation, far less is known on whether non-transplant treatment modes and especially high-flux dialysis also have a place in the therapy or prevention of dialysis-related amyloidosis. In this brief review we will examine whether effective therapeutic or preventive non-transplant strategies are available for dialysis-related amyloidosis, or whether, based on our fragmentary knowledge of the pathogenesis, such measures could be designed.

Pathogenesis of Dialysis-Related Amyloidosis

Retention and Synthetic Rates of β_2-Microglobulin in Chronic Renal Failure

β_2-Microglobulin is markedly retained in end-stage renal disease patients, since the kidney normally accounts for the elimination of about 95% of the normal β_2-microglobulin generation of 1,000–2,000 mg/week [9, 10]. Consequently, β_2-microglobulin levels in dialysis patients are elevated up to 60-fold [5]. This renal retention of β_2-microglobulin appears to be the basic requirement for the development of dialysis-related amyloidosis (fig. 1). In addition to the loss of renal function, the second determinant of the β_2-microglobulin accumulation in uremia is the β_2-microglobulin synthetic rate. One theory on the pathogenesis of dialysis-related amyloidosis suggests that β_2-microglobulin synthesis is stimulated by cytokines or other inflammatory stimuli induced during hemodialysis [1, 2, 11]. This is supported by in vitro studies showing that the cellular β_2-microglobulin synthesis rate can be enhanced by various inflammatory agents, including interleukin-1, tumor necrosis factor or lipopolysaccharide [12], or by direct contact of blood cells with certain dialysis membranes [13]. Further sup-

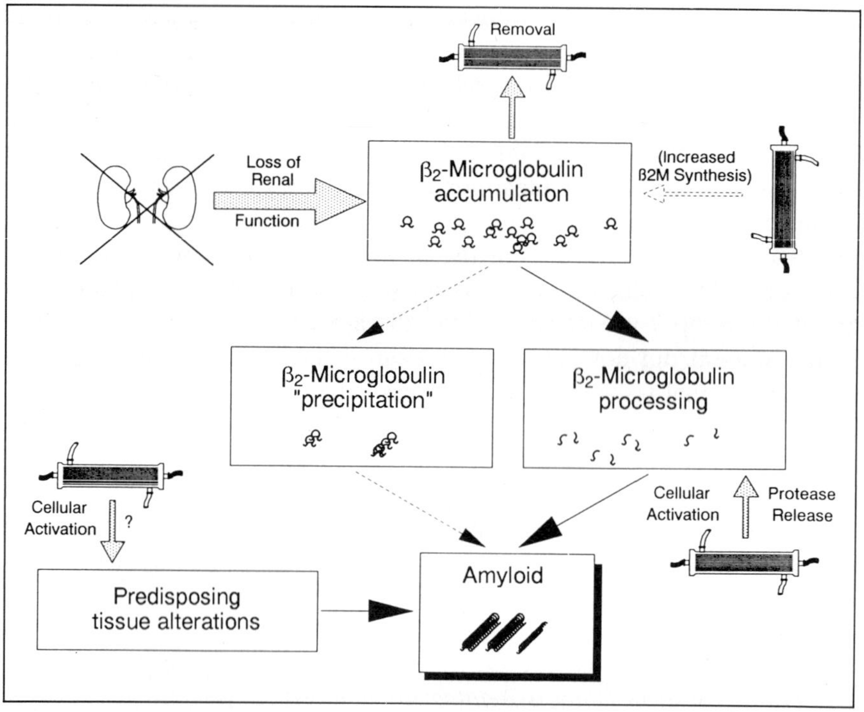

Fig. 1. Pathogenetic mechanisms possibly involved in the development of dialysis-related amyloidosis. Arrow widths indicate assumed importance of the respective factor, while dashed arrows indicate factors of unknown or questionable relevance.

port for this theory appeared to stem from the observation that with low-flux hemodialysis membranes a postdialytic increase of the β_2-microglobulin plasma level was observed, which could not be explained by mere hemoconcentration [11]. However, follow-up studies have demonstrated that these concentration changes largely represent complex water shifts between the extra- and intravascular space [14, 15] rather than a true increase of the β_2-microglobulin mass. This finding suggested that any stimulation of the β_2-microglobulin synthesis by various hemodialysis-related mechanisms observed in vitro is of a low order of magnitude in vivo, and that it cannot be detected by measurements of the β_2-microglobulin plasma concentration. To further investigate this issue, we performed in vivo studies with i.v.-injected radiolabelled β_2-microglobulin in

Table 1. Steady state plasma levels of endogenous β_2-microglobulin, metabolic clearance of i.v.-injected ^{131}I-β_2-microglobulin and the derived β_2-microglobulin synthetic rate (plasma level $\times$ metabolic clearance) in hemodialysis patients and in normal control persons: mean $\pm$ SD, data modified from Floege et al. [9]

	Plasma β_2-microglobulin mg/l	Metabolic clearance of ^{131}I-β_2-microglobulin ml/min	Synthetic rate of β_2-microglobulin	
			mg/week	mg/kg/day
Hemodialysis patients (n = 11)	37.6 ± 14.1	3.4 ± 0.7	1,491 ± 317	3.1 ± 0.8
Normal controls (n = 5)	1.9 ± 0.3*	65.5 ± 12.8*	1,271 ± 342	2.4 ± 0.7

* $p < 0.05$ vs. hemodialysis group.

order to determine its turnover [9]. For this the removal rates of endogenous β_2-microglobulin were determined indirectly by following the kinetics of the radioactive β_2-microglobulin. In a steady state condition (as defined by constant plasma levels) the removal rate of a molecule should equal its appearance rate, and hence these measurements allowed the estimation of the synthetic rates of β_2-microglobulin. As shown in table 1, no significant difference of the β_2-microglobulin synthetic rate was observed between normal controls and hemodialysis patients [9]. Although this study did not examine the possible influence of the different types of hemodialysis (high-flux versus conventional low-flux hemodialysis) on the β_2-microglobulin synthesis rate, our data indicated that the major determinant of the increased body burden of β_2-microglobulin was the loss of renal excretory function, while a hemodialysis-related enhancement of the β_2-microglobulin synthesis rate seems to be of comparatively low relevance in vivo (fig. 1).

Although the marked increase of the total body burden of β_2-microglobulin is likely to constitute the basis of the amyloidogenesis, and one study claimed that amyloid fibril-like structures may also be formed in vitro from intact β_2-microglobulin [16], two observations argue against a simple precipitation mechanism as the origin of AB-amyloid fibrils: (a) β_2-microglobulin plasma levels do not correlate with the presence or absence of disease [17], and (b) processed β_2-microglobulin molecules have been demonstated in the amyloid (see below) [18]. Therefore, further mechanisms appear to be operative in the pathogenesis of dialysis-related amyloidosis (fig. 1).

Cellular Activation and Processing of β_2-Microglobulin

A recent study on the dialysis-related amyloid fibrils demonstrated that in addition to intact β_2-microglobulin molecules, fragmented molecules, most of them with a lysine-specific cleavage, were also present [18]. Another study showed by two-dimensional electrophoresis that 'novel β_2-microglobulin' molecules with reduced molecular weights occurred in solubilized amyloid fibrils besides intact β_2-microglobulin [19]. Based on these findings it appears that a limited proteolysis of a circulating precursor molecule is involved in the development of dialysis-related amyloid deposits, which would be in accordance with current concepts on the pathogenesis of most types of amyloidosis. Therefore, although cellular activation during or after hemodialysis does not appear to play a major role in enhancing the β_2-microglobulin synthesis rate (see above), it may be very important in the context of β_2-microglobulin processing, and thus in the pathogenesis of dialysis-related amyloidosis (fig. 1). Such cellular activation, preceding the limited proteolysis of β_2-microglobulin, may take place during contact with the dialyzer membrane. Alternatively, resident or blood-derived cells in the tissue, especially macrophages, may be activated by mediators released from circulating cells during dialysis. Indeed, activation of cells with the subsequent release of mediators and/or proteases during or following hemodialysis has been widely documented [20–26].

Therapy of Dialysis-Related Amyloidosis

As with all other amyloidoses no curative therapy is known for an established dialysis-related amyloidosis and treatment consists strictly of attempting to relieve symptoms. In addition to preventing the further progression of dialysis-related amyloidosis, a successful renal transplantation has also been shown to result in rapid improvement of amyloid-associated symptoms like humeroscapular periarthritis [8]. It appears likely, however, that this relief of symptoms may have been due to the institution of an immunosuppressive regimen with suppression of the inflammatory reaction surrounding some amyloid deposits [27] rather than due to a regression of the deposits per se. No data are yet available on whether progression can also be prevented by changing from hemodialysis with β_2-microglobulin impermeable, low-flux membranes to hemodialysis with high-flux membranes. On the other hand, changing to treatment modes using high-flux membranes has been claimed to result in improvement of symptoms such as shoulder pain [28]. Whether this indeed is the case seems doubtful since there were no controls and the studies were not blinded, and objective parameters, if examined, failed to demonstrate any significant improvement [29]. Thus, these observations suggest that a change of treatment from using a low-flux to high-flux membrane is unlikely to have a substantial role in the therapy of an established amyloidosis.

Prevention of Dialysis-Related Amyloidosis by Treatment with High-Flux Membranes

Theoretical Considerations

High-flux membranes, such as hydrophilic polyamide, polysulfone and acrylonitrile, have β_2-microglobulin sieving coefficients of up to 0.6 and thereby allow greater removal compared to standard low-flux membranes [11, 30]. While some membranes, such as the acrylonitrile and the old hydrophobic polyamide version, allow some β_2-microglobulin removal via membrane adsorption [31], the main share of β_2-microglobulin removal during hemodialysis with high-flux membranes is achieved by diffusive forces [11, 30, 32]. The addition of convective forces, as generated with hemofiltration or to a lesser degree with hemodiafiltration, results in further augmentation of the β_2-microglobulin removal (fig. 2) [30]. Although β_2-microglobulin removal rates, similar to those of a normal kidney can be achieved with high-flux membranes, and assuming normal β_2-microglobulin synthesis rates in uremic patients [9, 10], the problem of β_2-microglobulin removal remains in the fact that 10–15 h treatment per week is minimal compared to 168 h of normal renal function per week. Accordingly, figure 3 shows that even a treatment with very high efficiency for β_2-microglobulin removal will not be able to normalize β_2-microglobulin plasma levels, as long as treatment time is not increased drastically. On the other hand, although a normalization of β_2-microglobulin levels to the normal range may not occur, it should be kept in mind that it is unknown so far whether a substantial reduction (e.g. 50%) of β_2-microglobulin plasma levels might in fact reduce the rate of amyloid formation. However,

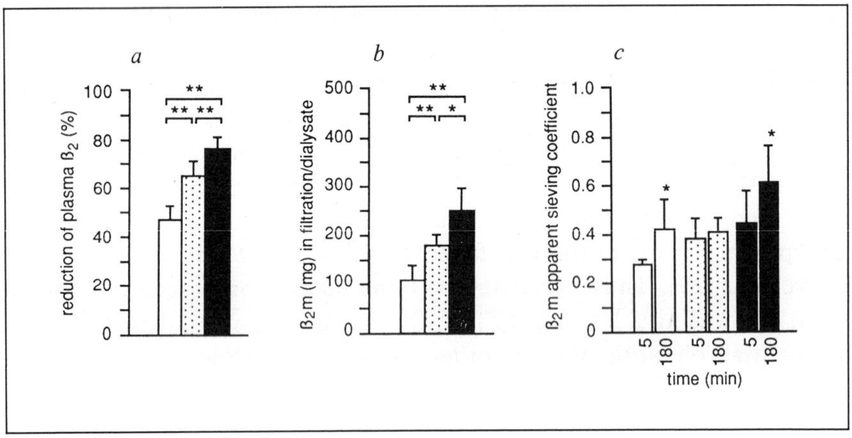

Fig. 2. β_2-microglobulin removal in HD (□), HDF (▩) and HF (■) with a hydrophilic polyamide dialyzer, Polyflux 160 (1.6 m^2). *a* Reduction of plasma level. *b* Recovery in ultrafiltrate/dialysate. *c* Apparent sieving coefficient after 5 and 180 min. In all cases mean ± SD of 6 experiments. * $p < 0.05$, ** $p < 0.01$. Data derived from Floege et al. [30].

as outlined above, blood levels of β_2-microglobulin do not correlate with the formation of amyloid, and other mechanisms besides β_2-microglobulin retention also appear to contribute to the development of amyloid deposits.

Blood contact with high-flux membranes, such as polyamide, polysulfone and acrylonitrile membranes, results in less granulocyte protease release, cytokine induction and complement activation, when compared to standard low-flux membranes [33–35]. Each of these mechanisms on a theoretical basis (see above) could have a beneficial influence on the β_2-microglobulin-derived amyloid formation. However, the possibility also exists that hemodialysis or hemodiafiltration with high-flux membranes could contribute to the disease process by facilitating the passage of endotoxin from dialysate to blood [36]. Endotoxin may also be implicated in the appearance of single cases of dialysis-related amyloidosis in rare patients exclusively treated on a long-term basis with high-flux membranes and hemofiltration [6]. This premise is based on the fact that in the past the pyrogen content of hemofiltration substitution fluid has not always been as low as today and (sub-)febrile reactions were a common occurrence during, for example, hemofiltration [37].

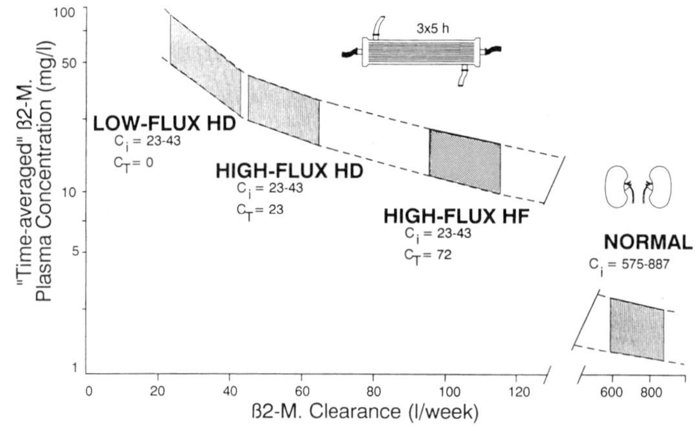

Fig. 3. Theoretical range of plasma β_2-microglobulin levels that can be achieved using either standard hemodialysis with β_2-microglobulin-impermeable membranes, or high-flux hemodialysis or hemofiltration with β_2-microglobulin-permeable membranes (β_2-microglobulin sieving coefficients about 0.6). Treatment time was assumed as 3×5 h/week for all treatment modes and ultrafiltration rates were assumed to be constant at about 15 ml/min in hemodialysis and 140 ml/min in hemofiltration. The calculations were based on the kinetic parameters for β_2-microglobulin determined previously [9]. For comparison similar estimates were performed for individuals with normal renal function (β_2-microglobulin sieving coefficient = 1.0, clearance 120 ml/min). The theoretical data thus obtained show a good agreement with β_2-microglobulin plasma levels measured in normals [5], patients on low-flux hemodialysis [5, 11] and the levels reported for patients on chronic high-flux hemodialysis with the AN 69 membrane [46]. C_i = Endogenous clearance of β_2-microglobulin; C_T = treatment derived clearance of β_2-microglobulin.

Retrospective Studies

Until now, the only retrospective data available are those comparing patients on long-term hemodialysis with Cuprophan® membranes with those on long-term hemodialysis with high-flux acrylonitrile (AN 69®) membranes. Other high-flux membranes have not been commercially available for a sufficient time to allow sufficient numbers of long-term patients to be studied. Data from a large multicenter study in Belgium and France comparing Cuprophan and AN 69 hemodialysis patients have suggested a lesser prevalence of dialysis-related amyloidosis in the high-flux group, as measured by the presence of carpal tunnel syndrome and/or cystic bone radiolucencies [38]. Similar conclusions were also reached in a smaller study in France [39]. In contrast, a recent EDTA survey [40] as well as another study in France [41] could not detect significantly different prevalences of amyloid-associated symptomatology in hemodialysis patients treated with either high-flux acrylonitrile or Cuprophan membranes. These discrepancies are likely to reside in the nature of retrospective data analysis with variables like water purity, age at the onset of dialysis treatment, membrane characteristics, reuse policies and dialysate composition being largely uncontrolled and/or changing over time. Also, very few patients of these studies had been treated consistently with the same membrane, and all studies have accepted patients for the high-flux group who had received up to 10–20% of their total hemodialysis time with low-flux hemodialysis or by other treatment modes. Finally, all retrospective studies hitherto were based solely on clinical and/or radiologial findings, which may have multifactorial origins and also are the least sensitive indicators of dialysis-related amyloidosis [42]. Thus, it is clear that prospective studies and more reliable clinical amyloid markers will be necessary to evaluate the issue of prevention of dialysis-related amyloidosis and high-flux membranes or high-flux dialysis.

Prospective Studies and How to Control Their Efficiency

In the design of a prospective study examining the influence of high-flux membranes on the development of dialysis-related amyloidosis, the main concern currently would be related to contaminated dialysate (see above). Therefore, the use of pyrogen-free dialysate should be mandatory in such studies. Ideally, such a study should compare patients on standard low-flux hemodialysis with those on hemofiltration with biocompatible high-flux membranes, since the latter would offer the least cell activation and protease release as well as the maximal β_2-microglobulin removal currently achievable.

The prospective evaluation of such a study hitherto would have suffered from the non-availability of an early, easily accessible and sensitive parameter to indicate the presence of amyloid anywhere in the body. Amyloid-related clinical symptomatology or radiological findings are late occurrences in the disease process and biopsies of non-selected tissues may at best yield chance findings of amyloid [1, 2, 42]. The two procedures, which most often seem to yield positive results, i.e. a synovial biopsy or an aspiration of a joint effusion, respectively [42], certainly do not have the poten-

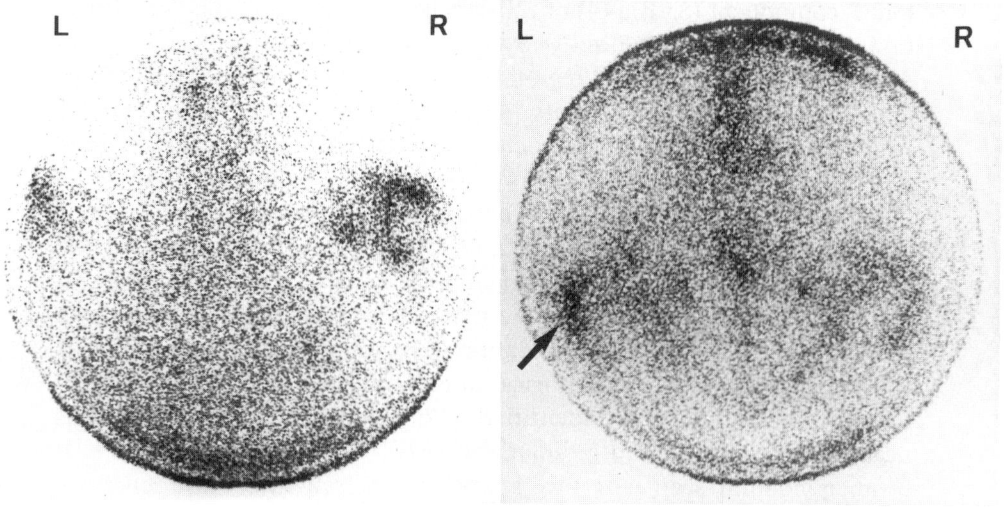

a *b*

Fig. 4. Representative images obtained after in vivo injection of ^{131}I-β_2-microglobulin. Reprinted with permission from Floege et al. [45]. *a* Dorsal image of the shoulders and neck of a patient in whom clinical and radiological signs of a left-sided humeroscapular periarthiritis were present. *b* Dorsal image of the pelvic area of a patient showing local accumulation in both hip regions, pelvic sites and over the lumbar vertebral column. Clinically and radiologically a cystic radiolucency (biopsy-proven AB amyloid) of the left neck of the femur was noted.

tial to become routine diagnostic procedures. While the plasma concentration of β_2-microglobulin has repeatedly been shown to be of no diagnostic value [17], the occurrence of cleaved or 'novel' β_2-microglobulin [19] might become a marker of incipient or ongoing amyloid deposition, but no assays for these β_2-microglobulin species are yet available. Given these constraints it appeared that a diagnostic test in which the amyloid deposits are (radio-)labelled and then visualized would be an ideal way to screen for dialysis-related amyloidosis and/or to follow the course of the amyloidosis.

Initial attempts to visualize amyloid deposits with radiolabelled diphosphonates by virtue of their high calcium content have been hampered by the non-specificity of this method [43]. However, recently two promising reports have appeared, describing the diagnostic imaging of dialysis-related amyloid deposits by i.v. injection of either ^{123}I-labelled serum amyloid P component (SAP) [44] or ^{131}I-labelled β_2-microglobulin [45]. Injection of radiolabelled SAP allowed the non-invasive detection of amyloid deposits in the carpal tunnel region and metacarpophalangeal joints of two long-term hemodialysis patients [44], but it has not yet been reported whether this scan is also able to detect asymptomatic, early stages of amyloid deposition. An alternative is offered by the ^{131}I-β_2-microglobulin scan [45]. Using this technique, it has been possible to image dialysis-related amyloid deposits in the shoulders, knees, pelvis, hands and vertebral column of long-term hemodialysis patients with clinical correlates of dialysis-related amyloidosis (fig. 4). Furthermore, this study showed that about 50% of the sites of tracer accumulation were not associated with any clinical symptomatology or radiological alterations, and in fact several amyloid-positive patients were identified who would have been classified as negative on a clinical and radiological basis [45]. Consequently, the ^{131}I-β_2-microglobulin scan currently seems to represent the most sensitive, objective marker of dialysis-related amyloidosis and by repeated scanning, it should permit the prospective evaluation of the influence of high-flux dialysis on the appearance or progression of amyloid deposits.

References

1 Kleinman KS, Coburn JW: Amyloid syndromes associated with hemodialysis. Kidney Int 1989;35:567–575.
2 Bardin T, Zingraff J, Kuntz D, Drüeke T: Dialysis-related Amyloidosis. Nephrol Dial Transplant 1986;1:151–154.
3 Charra B, Calemard E, Laurent G: Chronic renal failure treatment duration and mode: their relevance to the late dialysis periarticular syndrome. Blood Purif 1988; 6:117–124.
4 Gejyo F, Odani S, Yamada T, Honma N, Saito H, Suzuki Y, Nakagawa Y, Kobayashi H, Maruyama Y, Hirasawa Y, Suzuki M, Arakawa M: β_2-Microglobulin: A new form of amyloid protein associated with chronic hemodialysis. Kidney Int 1986;30: 385–390.
5 Vincent, C, Revillard JP, Galland M, Traeger J: Serum β_2-microglobulin in hemodialyzed patients. Nephron 1978;21:260–268.
6 Renaud H, Fournier A, Moriniere P, Westeel PF, El Esper N, Belbrik S, Marie A, Cohen-Solal HE, Selbert JL: Erosive osteoarthropathy associated with β_2-microglobulin in a uremic patient treated exclusively by long-term hemofiltration with biocompatible membranes. Nephrol Dial Transplant 1988;3:820–822.
7 Gagnon RF, Somerville P, Kaye M: β_2-Microglobulin serum levels in patients on long-term dialysis. Periton Dial Bull 1987;7:29–31.

8 Jadoul M, Maldague M, Pirson Y, Van Ypersele de Strihou C: Renal transplantation for dialysis arthropathy (letter). Lancet 1988;ii:849.
9 Floege J, Bartsch A, Schulze M, Shaldon S, Koch KM, Smeby LC: Turnover of ^{131}I-β_2-microglobulin in hemodialyzed patients. J Lab Clin Med 1991;118:153–165.
10 Odell RA, Slowiaczek P, Moran JE, Schindhelm K: Beta$_2$-microglobulin kinetics in end-stage renal failure. Kidney Int 1991;39:909–919.
11 Floege J, Granolleras C, Bingel M, Deschodt G, Branger B, Oules R, Koch KM, Shaldon S: β_2-Microglobulin kinetics during haemodialysis and haemofiltration. Nephrol Dial Transplant 1987;1:223–228.
12 Knudsen PJ, Leon J, Ng AK, Shaldon S, Floege J, Koch KM: Hemodialysis-related induction of β_2-microglobulin and interleukin-1 synthesis and release by mononuclear phagocytes. Nephron 1989;53:188–193.
13 Zaoui PM, Stone WJ, Hakim RM: Effects of dialysis membranes on β_2-microglobulin production and cellular expression. Kidney Int 1990;38:962–968.
14 Floege J, Granolleras C, Merscher S, Eisenbach GM, Colton CK, Koch KM, Shaldon S: Is the rise in plasma β_2-microglobulin seen during hemodialysis meaningful? Nephron 1989;51:6–12.
15 Bergström J, Wehle B: No change in corrected β_2-microglobulin concentration after cuprophane haemodialysis (letter). Lancet 1987;i:628–629.
16 Connors LH, Shirahama T, Skinner M, Fenves A, Cohen AS: In vitro formation of amyloid fibrils from intact β_2-microglobulin. Biochem Biophys Res Commun 1989;131:1063–1068.
17 Gejyo F, Homma N, Suzuki Y, Arakawa M: Serum levels of β_2-microglobulin as a new form of amyloid protein in patients undergoing long-term hemodialysis (letter). N Engl J Med 1986;314:585–586.
18 Linke RP, Hampl H, Lobeck H, Ritz E, Bommer J, Waldherr R, Eulitz M: Lysine-specific cleavage of β_2-microglobulin in amyloid deposits associated with hemodialysis. Kidney Int 1989;36:675–681.
19 Ogawa H, Saito A, Oda O, Nakajima M, Chung TG: Detection of novel β_2-microglobulin in the serum of hemodialysis patients and its amyloidogenic predisposition. Clin Nephrol 1988;30:158–163.
20 Hörl WH, Schaefer RM, Heidland A: Effect of different dialyzers on proteinases and proteinase inhibitors during hemodialysis. Am J Nephrol 1985;5:320–326.
21 Lonnemann G, Bingel M, Koch KM, Shaldon S, Dinarello CA: Plasma interleukin-1 activity in humans undergoing hemodialysis with regenerated cellulosic membranes. Lymphokine Res 1987;6:63–70.
22 Haeffner-Cavaillon N, Cavaillon JM, Cianocioni C, Bacle F, Delons S, Kazatchkine MD: In vivo induction of interleukin-1 during hemodialysis. Kidney Int 1989;35:1212–1218.
23 Hakim RM, Breillatt J, Lazarus JM, Port FK: Complement activation and hypersensitivity reactions to dialysis membranes. N Engl J Med 1984;311:878–882.
24 Lonnemann G, Bingel M, Floege J, Koch KM, Shaldon S, Dinarello CA: Detection of endotoxin-like interleukin-1 inducing activity during in vitro dialysis. Kidney Int 1988;33:29–35.
25 Bingel M, Lonnemann G, Koch KM, Shaldon S: Enhancement of in-vitro human interleukin-1 production by sodium acetate. Lancet 1987;i:14–16.

26 Thu Nguyen A, Lethias C, Zingraff J, Herbelin A, Naret C, Descamps-Latscha B: Hemodialysis membrane-induced activation of phagocyte oxidative metabolism detected in vivo and in vitro within microamounts of whole blood. Kidney Int 1985; 28:158–167.

27 Coggi G, Maiorano E, Braidotti P, Dell'Orto P, Viale G: Beta-2-microglobulin amyloidosis in synovial membranes: Histopathology, immunocytochemistry and electron microscopy; in Gejyo F, Brancaccio D, Bardin T (eds): Dialysis Amyloidosis. Milano, Wichtig Editore, 1989, pp 57–78.

28 Hardouin P, Flipo RM, Foissac-Gegoux P, Dumont A, Duquesnoy B, Delcambre B: Dialysis related beta-2-microglobulin amyloid arthropathy. Improvement of clinical symptoms after a switch of dialysis membranes. Clin Rheumatol 1988;7:41–45.

29 Sethi D, Brown EA, Perkin GD, Maini RN, Gower PE: High-flux hemodialysis: optimal treatment for dialysis arthropathy? XXVIth Congr European Dialysis and Transplant Association, Göteborg, 1989, abstr book, p 220.

30 Floege J, Granolleras C, Deschodt G, Heck M, Baudin G, Branger B, Tournier O, Reinhard B, Eisenbach GM, Smeby LC, Shaldon S: High-flux synthetic versus cellulosic membranes for β_2-microglobulin removal during hemodialysis, hemodiafiltration and hemofiltration. Nephrol Dial Transplant 1989;4:653–657.

31 Floege J, Granolleras C, Smeby LC; Göhl H, Koch KM, Shaldon S: Hydrophilic high flux polyamide membranes for β_2-microglobulin removal. Trans Am Soc Artif Organs 1987;32:309–311.

32 Mineshima M, Hoshino T, Era K: Diffusive and convective mass transport characteristics of β_2-microglobulin removal. Trans Am Soc Artif Organs 1987;33:103–106.

33 Hörl WH, Steinhauer HB, Riegel W, Schollmeyer P: Plasma granulocyte elastase during hemodialysis: Effects of different dialyzer membranes and sterilization procedures. Proc Eur Dial Transplant Assoc 1985;22:212–215.

34 Levett DL, Woffindin C, Bird AG, Hoenich NA, Ward MK, Kerr DNS: Complement activation in hemodialysis: A comparison of new and re-used dialyzers. Int J Artif Organs 1986;9:97–104.

35 Lonnemann G, Koch KM, Shaldon S, Dinarello CA: Studies on the ability of hemodialysis membranes to induce, bind, and clear human interleukin-1. J Lab Clin Med 1988;112:76–86.

36 Lonnemann G, Behme TC, Lenzner B, Floege J, Mahiout A, Colton CK, Koch KM, Shaldon S: The permeability of dialyzer membranes to TNF_α inducing substances derived from water bacteria. Submitted.

37 Frei U, Koch KM: Fever and shock during hemofiltration. Contr Nephrol. Basel, Karger, 1983, vol 36, pp 107–114.

38 Van Ypersele C, Jadoul M, Malghem J, Maldague B, Jamart J, and Working Party on Dialysis Amyloidosis: Effect of dialysis membrane and patient's age on signs of dialysis-related amyloidosis. Kidney Int 1991;39:1012–1019.

39 Lavaud S, Toupance O, Roujouleh H, Fakir M, Melin JP, Chanard J: Carpal tunnel syndrome in hemodialysis patients: Effect of dialysis strategy; in Man, Mion, Henderson (eds): Blood Purification in Perspective: New Insights and Future Trends. Cleveland, ISAO Press, 1987, pp 125–128.

40 Brunner FP, Brynger H, Ehrich JHH, Fassbinder W, Geerlings W, Rizzoni G, Selwood NH, Tufveson G, Wing AJ: Case control study on dialysis arthropathy: The

influence of two different dialysis membranes: data from the EDTA registry. Nephrol Dial Transplant 1990;5:432–436.
41 Kessler M, Netter P, Maheut H, Wolf C, Prenat E, Cao Huu T, Gaucher A: Highly permeable and biocompatible membranes and prevalence of dialysis associated arthropathy (letter). Lancet 1991;337:1092.
42 Floege J: Diagnostic approaches to AB-amyloidosis. Blood Purif, in press.
43 Sethi D, Naunton Morgan TC, Brown EA, Gower PE: Technetium-99-labelled methylene diphosphonate uptake scans in patients with dialysis arthropathy. Nephron 1990;54:202–207.
44 Hawkins PN, Myers MJ, Lavender JP, Pepys MB: Diagnostic radionuclide imaging of amyloid: Biological targeting by circulating human serum amyloid P component. Lancet 1988;i:1413–1418.
45 Floege J, Burchert W, Brandis A, Gielow P, Nonnast-Daniel B, Spindler E, Hundeshagen H, Shaldon S, Koch KM: On the imaging of dialysis related amyloid (AB-amyloid) deposits with ^{131}I-β_2-microglobulin. Kidney Int 1990;38:1169–1176.
46 Zingraff J, Beyne P, Urena M, Uzan M, Nguyen Khoa M, Descamps-Latscha B, Drüeke T: Influence of hemodialysis membranes on β_2-microglobulin kinetics: In vivo and in vitro studies. Nephrol Dial Transplant 1988;3:284–290.

Dr. Jürgen Floege, Abteilung Nephrologie 6840, Medizinische Hochschule,
D-W-3000 Hannover 61 (FRG)

Shaldon S, Koch KM (eds): Polyamide – The Evolution of a Synthetic Membrane for Renal Therapy. Contrib Nephrol. Basel, Karger, 1992, vol 96, pp 138–144

Removal of High-Molecular-Weight Toxins

W.H. Hörl

Department of Medicine, Division of Nephrology, University of Saarland, Homburg/Saar, FRG

High-molecular-weight solutes are not removed by hemodialysis with conventional membranes. For this purpose highly permeable membranes, such as the polyamide, the polysulfone and the AN69, and therapies involving convective transport, such as hemofiltration and hemodiafiltration are needed. The removal of a certain substance by ultrafiltration depends on its blood concentration, on the ultrafiltration flow rate and on the membrane sieving coefficient. The sieving properties may vary due to both nonspecific adsorption of plasma proteins and specific adsorption of the studied solute [1, 2].

To mimic the function of the glomerular basement membrane, chemical engineering has produced more and more open membranes. This trend has also been influenced by the recognition of large molecules, such as PTH and β_2-microglobulin, as potential uremic toxins, and the belief that their removal would be therapeutically beneficial. Other large uremic solutes are also being identified and characterised, as will be discussed below. Long-term experience with these membranes has already shown that depletion syndromes do not seem to occur. Further experience, particularly with prospective studies, are still needed to determine if these membranes have a clear beneficial effect on uremic complications.

Parathyroid Hormone

One of the uremic toxins involved in uremic complications, such as bone disease, anemia, lipid disorders, or polyneuropathy is parathyroid hormone (PTH). Various reports indicate that a single hemofiltration session induces a significant decrease in serum PTH concentration. Schaefer et al. [3] found that the fall of serum PTH during hemofiltration was not only related to treatment induced changes in serum calcium concentration, but was also caused by removal of PTH. There is clear evidence that, in addition to biologically inactive fragments of PTH, the intact hormone is removed during hemofiltration treatment [4]. Some authors found that the serum PTH concentration returned to normal a few hours after the hemofiltration treatment [5]. However, Schulz et al. [6] observed lower PTH values in hemofiltration patients as compared to patients undergoing conventional hemodialysis, but there were no significant changes of the underlying bone disease. Long-term observations of hemoglobin values, various lipid disorders and nerve conduction velocity in patients treated with hemofiltration have not shown significant improvement [4].

Granulocyte Inhibitory Protein

Granulocyte inhibitory protein (GIP) was isolated from ultrafiltrate obtained from chronically uremic patients dialyzed with high-flux membranes. The filtrate was concentrated 25-fold by means of an Amicon ultrafiltrate cell with a PM 10 membrane, which has a molecular weight cut-off of approximately 10,000 daltons. The concentrate was dialyzed and then chromatographed on an ion-exchange fast protein liquid chromatographic (FPLC) system (Pharmacia, Uppsala, Sweden). The material was eluted with a linear NaCl gradient. The fractions that contained biological activity were pooled and were concentrated by lyophilization. The material was resuspended in water and chromatographed with a high-performance size-exclusion column (Zorbax GF 250, Du Pont de Nemours, Homburg, FRG). The purification process concentrated the protein 240-fold and to a purity of $>95\%$. The polypeptide had a molecular weight of approximately 28,000 daltons and an isoelectric point of 4.0–4.5. The polypeptide inhibited the uptake of deoxyglucose, chemotaxis, oxidative metabolism, and intracellular bacterial killing by polymorphonuclear leukocytes (PMNL). The IC_{50} of this granulocyte inhibitory protein required to inhibit both the biological and the functional changes of PMNL is in the nanomolar range, calculated on the basis of a molecular weight of 28,000 daltons for GIP. Therefore, the efficacy of the protein is well within the range of physiological effector substances (table 1). A specific rabbit polyclonal antibody raised against the protein nullified these inhibitory

Table 1. Biological activity of the granulocyte inhibitory protein

Test condition	GIP concentration which inhibits 50% of maximal stimulation (IC_{50}), mg/ml
^{3}H *d*-glucose uptake[1]	5.2 ± 1.2 (n = 5)
Chemotaxis[1]	7.2 ± 1.7 (n = 6)
Oxidative metabolism[1]	8.1 ± 1.4 (n = 5)
Intracellular killing[2]	7.9 ± 1.7 (n = 6)

[1] Stimulation with FMLP.
[2] Stimulation with *S. aureus* 502.

changes. Amino acid sequencing of the NH_2-terminus revealed a single sequence that proved to be nonhomologous with other serum proteins that appear during an inflammatory state, which include endotoxins, interleukins, interferons, complement factors, colony-stimulating factor, and tumor necrosis factor α and β [7].

GIP could be isolated from ultrafiltrate from all hemodialysis patients tested using the polyamide or the polysulfone membrane. The amount of GIP obtained was comparable with both membranes. None of our patients had any sign of infection. Therefore, it is unlikely that GIP is derived from infecting microorganisms. Extending the computer search we found some homology with a paraprotein (molecular weight 10,000 daltons) isolated by Carson and co-workers [8, 9]. This protein is found in 2% of circulating IgM from normal adults, approximately 40% of circulating rheumatoid factors from some patients with Sjogren's syndrome and may be associated with certain B cell lymphoproliferative diseases. It appears paradoxical that a circulating factor (e.g. immunoglobulin fragment) has an inhibitory effect on the host's cellular defense mechanism under biological conditions. However, since the biological activity of PMNL can be restored after the removal of GIP, it is possible that the function of this protein is of physiological relevance. Baum et al. [10] demonstrated that chemotaxis of PMNL was impaired in the presence of undialyzed uremic sera, but that hemodialysis seemed to correct the defect. A chemotactic inhibitor found in sera of uremic patients acted directly on leukotactic factors (C3 and C5 chemotactic fragment and bacterial factor) to render them irreversibly inactive [11]. Pedersen et al. [12] studied the ability of serum to attract PMN in uremic patients before and after hemodialysis. The chemotactic response toward serum from uremic patients was significantly decreased prior to hemodialysis and was normalized thereafter. Phagocytosis in uremic patients is either normal or slightly diminished. Hirabayashi et al. [13] observed decreased phagocytic uptake of IgG-coated particles in patients before hemodialysis therapy, which was restored by hemodialysis. Kinetic measurements of the serum-independent uptake of IgG-coated polyvinyl toluene latex particles by isolated PMNL cells have been performed in patients undergoing hemodialysis or peritoneal dialysis. The mean phagocytic rate for the patient group was significantly reduced when compared to the reference group of apparently healthy individuals. After four months of adequate dialysis, the phagocytic uptake was significantly higher than the uptake shortly after the beginning of dialysis [14].

Degranulation Inhibiting Protein

Ultrafiltrate obtained from chronically uremic patients dialyzed with high-flux membranes was concentrated 300-fold by means of an Amicon ultrafiltration cell using a YM5 membrane (Amicon, Ousterhous, USA) with a cut-off of approximately 5,000 daltons. Crude concentrated ultrafiltrate was chromatographed on an Ultrogel AcA54 gel-permeation column. Fractions showing degranulation inhibiting activity were concentrated, dialyzed and applied to a Q-Sepharose fast flow column. Degranulation inhibiting protein (DIP) was pooled, concentrated and loaded onto a blue Sepharose column. Bound protein was eluted using a linear NaCl gradient. Biologically active protein was dialyzed and chromatographed on a Mono S HR 5/5 column (Pharmacia, Uppsala, Sweden). DIP was eluted using a linear NaCl gradient [15]. Analysis for DIP purity by sodium dodecyl sulfate-polyacrylamide gel electrophoresis (SDS/PAGE) and high-performance liquid chromatography (HPLC) showed that the protein was homogenous and free from contaminating protein within the detection limits of the methods employed. The molecular weight of DIP was determined to be approximately 14,000 daltons (nonreduced SDS/PAGE). Automatic Edman degradation of the protein was unsuccessful indicating a blocked N-terminus. After BrCN cleavage, sequencing of DIP provided a single partial sequence of 16 amino acids which was identical with the angiogenin residues 30–45. This angiogenesis factor has been reported to have a molecular weight of 14,400 daltons and a blocked N-terminus [16]. Angiogenin is not only a tumor-derived protein, but also a plasma protein [17] which quickly binds to the membrane of different cell lines. Angiogenin has the ability to intervene with cellular signal transduction [18]. The protein activates the phosphoinositide-specific phospholipase C in cultured rat aortic smooth muscle cells [19]. Exposure of granulocytes to DIP resulted in a depression of lactoferrin, collagenase and gelatinase secretion. Secretion of elastase, induced by cytochalasin B, was not depressed. Phagocytosis, oxidative respiratory burst and chemotaxis of the granulocytes were not affected by DIP. Recombinant angiogenin induced the same cellular effects in PMNL as DIP [15].

Several angiogenic factors have been fully purified, their amino acid sequences determined, and their genes cloned. These polypeptides include acidic and basic fibroblast growth factor, angiogenin, and transforming growth factors α and β. Angiogenic factors can be subdivided into two groups: (a) those that act directly on vascular endothelial cells to stimulate locomotion or mitosis, and (b) those that act indirectly by mobilizing host cells to release endothelial growth factors [20]. A protein produced by human colon cancer cells was isolated and characterized by Vallee and his collegues [21] and named angiogenin. The protein induces neo-vascularization on the chick chorioallantoic membrane at femtomolar levels and in the rabbit cornea at picomolar concentrations. A protein which is physically and functionally identical with angiogenin is present in normal human plasma and was purified to homogeneity by Shapiro et al. [17]. The possible implication of angiogenic factors in uremia needs to be investigated.

Conclusions

It has been demonstrated that the polyamide membranes provide high hydraulic and diffusive permeability for both small and large molecular substances. Use of the membrane led to effective elimination of a wide range of uremic toxins. The importance of high-molecular-weight uremic toxins is illustrated by PTH and β_2-microglobulin and by the recently purified and characterised GIP and DIP. It is very likely that there are a number of as yet unknown, large, uremic toxins which remain to be identified. Highly permeable membranes, like polyamide, will play an important role both in isolating these substances and possibly in removing them from the blood therapeutically. Monoclonal antibodies against GIP, DIP and other as yet unidentified factors will allow for characterizing their biological role.

Furthermore, monoclonal antibodies against perturbing factors may alleviate the abnormal responsiveness at the cellular level seen in chronically uremic patients.

References

1 Röckel A, Hertel J, Fiegel P, Abdelhamid S, Panitz N, Walb D: Permeability and secondary membrane formation of a high flux polysulfone hemofilter. Kidney Int 1986;30:429–432.

2 Chanard J, Toupance O, Gillery P, Lavaud S: Evaluation of protein loss during hemofiltration. Kidney Int 1988;33:S114–S116.

3 Schaefer K, Offermann G, von Herrath D, Asmus G, Hüfler M: Parathyroid hormone, 25-OH-vitamin D, and digoxin levels in patients treated by chronic hemofiltration. J Dial 1977;1:619–630.

4 Schaefer K, von Herrath D, Pauls A, Hüfler M: New insight into hemofiltration. Semin Nephrol 1986;6:161–167.

5 Pierides AM, Giacherio D, Schniepp B, Burritt M: Calcium, phosporus and parathyroid hormone metabolism in chronic hemofiltration. Contrib Nephrol. Basel, Karger, 1982, vol 32, pp 77–85.

6 Schulz W, Baier E, Hümpfner A, Delling G: Clinical problems of renal osteopathy in patients on hemofiltration. Contrib Nephrol. Basel, Karger, 1982, vol 32, pp 86–91.

7 Hörl WH, Haag-Weber M, Georgopoulos A, Block LH: Physicochemical characterization of a polypeptide present in uremic serum that inhibits the biological activity of polymorphonuclear cells. Proc Natl Acad Sci USA 1990;87:6353–6357.

8 Jirik FR, Sorge J, Fong S, Heitzmann IG, Curd IG, Chen PP, Goldfien R, Carson DA: Cloning and sequence determination of a human rheumatoid factor light-chain gene. Proc Natl Acad Sci USA 1986;83:2195–2199.

9 Silverman GJ, Goldfien RD, Chen P, Mageed RA, Jefferis R, Goni F, Frangione B, Fong S, Carson DA: Idiotypic and subgroup analysis of human monoclonal rheumatoid factors. Implications for structural and genetic basis of autoantibodies in humans. J Clin Invest 1988;82:469–475.

10 Baum J, Cestero RVM, Freeman RB: Chemotaxis of polymorphonuclear leukocytes and delayed hypersensitivity in uremia. Kidney Int 1975;7:S147–S153.

11 Siriwatratananonta P, Sinsakul V, Stern K, Slavin RG: Defective chemotaxis in uremia. J Lab Clin Med 1978;92:403–407.

12 Pedersen JO, Knudsen F, Nielsen AH, Grunnet N: The ability of uremic serum to induce neutrophil chemotaxis in relation to hemodialysis. Blood Purif 1987;5:24–28.

13 Hirabayashi Y, Kobayashi T, Nishikawa A, Aoki T, Takaya J, Kobayashi Y: Oxidative metabolism and phagocytosis of polymorphonuclear leukocytes in patients with chronic renal failure. Nephron 1988;49:305–312.

14 Hällgren R, Fjellström KE, Venge P: Kinetic studies of phagocytosis. II. The serum-independent uptake of IgG-coated particles by polymorphonuclear leukocytes from uremic patients on regular dialysis treatment. J Lab Clin Med 1979;94:277–284.

15 Hempelmann U, Reinke H, Hörl WH, Tschesche H: Inhibition of degranulation of polymorphonuclear leukocytes by a protein identical to angiogenin. Eur J Biochem, in press.
16 Kurachi K, Davie EW, Strydom DJ, Riordan JF, Vallee BL: Sequence of the cDNA and gene for angiogenin, a human angiogenesis factor. Biochemistry 1985;24:5494–5499.
17 Shapiro R, Strydom DJ, Olson KA, Vallee BL: Isolation of angiogenin from normal human plasma. Biochemistry 1987;26:5141–5146.
18 Heath WF Jr, Moore F, Bicknell R, Vallee BL: Modulation of mitogenic stimuli by angiogenin correlates with in vitro phosphatidylinositol biophosphate synthesis. Proc Natl Acad Sci USA 1989;86:2718–2722.
19 Moore F, Riordan JF: Angiogenin activates phospholipase C and elicts a rapid incorporation of fatty acid into cholesterol esters in vascular smooth muscle cells. Biochemistry 1990;29:228–233.
20 Folkman J, Klagsbrun M: Angiogenic factors. Science 1987;235:442–447.
21 Fett JW, Strydom DJ, Lobb RR, Alderman EM, Bethune JL, Riordan JF, Vallee BL: Isolation and characterization of angiogenin, and angiogenic protein from human carcinoma cells. Biochemistry 1985;24:5480–5486.

Prof. Dr. W.H. Hörl, Medizinische Universitätsklinik IV,
D-W-6650 Homburg/Saar (FRG)

Subject Index